WEIRDER MATHS

ABOUT THE AUTHORS

David has a PhD in astronomy from the University of Manchester. For the past 35 years he has been a freelance science writer and is the author of nearly 50 books on subjects such as cosmology, physics, philosophy, and mathematics. His website, The Worlds of David Darling, www.daviddarling.info, has been a widely used online resource for the past 20 years. He also tutors students in maths and physics and this is how he first met the 13-year-old Agnijo.

Agnijo was born in Kolkata, India, but has spent most of his life in Scotland. His extraordinary mathematical talents were recognised at an early age. In 2018 he came joint first in the International Mathematical Olympiad, recording a perfect score and affirming his status as one of the world's most outstanding young mathematicians. He is now continuing his studies at Trinity College, Cambridge.

WEIRDER MATHS

At the Edge of the Possible

DAVID DARLING
AND
AGNIJO BANERJEE

ONEWORLD

A Oneworld Book

First published by Oneworld Publications, 2019

Copyright © David Darling and Agnijo Banerjee 2019

The moral right of David Darling and Agnijo Banerjee to be
identified as the Authors of this work has been asserted by them in
accordance with the Copyright, Designs, and Patents Act 1988

A CIP record for this title is available from the British Library

ISBN 978-1-78607-508-6
eISBN 978-1-78607-509-3

Illustrations: Chartres-style labyrinth © Luca Galuzzi; Jubilee maze ©
NotFromUtrecht (Wikimedia); Mammoth Cave © United States Geological
Survey; Clay tablet © BabelStone (Wikimedia); Torricelli experimenting and
Niccolò Tartaglia © Wellcome Collection; Quartz crystals © JJ Harrison; Cosmic
background radiation map © NASA/WMAP Science Team; The Doryphoros ©
Ricardo André Frantz; Florence cathedral dome © Florian Hirzinger; Semi-regular
tiling © R. A. Nonenmacher; Alhambra tiling © Kolforn (Wikimedia); Ramanujan
bust © AshLin (Wikimedia); Giant soap bubble © Kazbeki (Wikimedia); T-Puzzle
© Voorlandt (Wikimedia); Reuleaux triangle © Frédéric Michel; Lovell telescope
© Mike Peel (Jodrell Bank Centre for Astrophysics, University of Manchester)

Typeset by Tetragon, London
Printed and bound in Great Britain by Clays Ltd, Elcograf S.p.A.

Oneworld Publications
10 Bloomsbury Street
London WC1B 3SR
England

In the broad light of day mathematicians check their equations and their proofs, leaving no stone unturned in their search for rigour. But, at night, under the full moon, they dream, they float among the stars and wonder at the miracle of the heavens. They are inspired. Without dreams there is no art, no mathematics, no life.

– Michael Atiyah

In mathematics you don't understand things. You just get used to them.

– John von Neumann

Contents

Introduction

THIS IS OUR second foray into some of the most outrageous, fascinating, and downright peculiar parts of mathematics, following our earlier adventures in *Weird Maths*. We'll be venturing into a land of bizarre shapes and numbers, exploring, like Gulliver, realms of the ultra small and the fantastically large, wandering down passageways with many twists and turns, and, along the way, coming face to face with some of the greatest challenges the human mind has ever confronted.

Maths is a subject vaster than most of us realise – so scarily vast that perhaps we're lucky to have minds limited in how far they can see. Maths penetrates every aspect of our lives and underpins not just science and technology, but also music and the arts, the forms, patterns, and movements that surround us, and even the games we play. It can be as hard as solving a page full of convoluted equations in a graduate class at Princeton or as easy as blowing bubbles for a child. We do maths every minute of every day because maths infuses all aspects of the universe around us, part of the infrastructure of reality. Some of it's as familiar as 1, 2, 3, or the symmetry of a circle. But much of maths is extraordinary and dazzling, beautiful, diverse, and strange. Its wonder and weirdness literally have no bounds.

As a writing team we're a little unusual. One of us (David) is a physicist and astronomer by training, who's spent the past 35 years writing books on everything from cosmology to consciousness. The other (Agnijo) is a teenage maths prodigy who started coming for private lessons with David several years ago and in 2018 took joint first place in the International Mathematical Olympiad with a perfect score of 42 points out of 42. He recently headed off to Cambridge University to continue his mathematical explorations there. About three years ago we began work on *Weird Maths*, dividing the chapters between us, and then cross checking each other's work – Agnijo focusing on the maths itself while David concentrated on making the writing clear and adding historical and biographical detail. The collaboration proved so successful and relatively stress-free that here we are again with the sequel!

So much is happening in this field – the pace of discovery is so dizzying – that we've added more chapters to *Weirder Maths*. Our goals, however, remain the same. We want to bring the most unusual, interesting, and important ideas in maths within reach of the general reader, not avoiding topics simply because they may seem hard to explain. Our mantra continues to be that anyone can grasp maths if it's framed in the right language. We've also tried, whenever possible, to show how the maths is relevant in everyday life or how it's proved useful to science and other fields.

We hope that some of the enthusiasm we feel for this most amazing and often misunderstood subject spills over onto these pages. Maths truly can be weird but, more than anything, it's a very human endeavour, full of the fun and foibles that mark us out as a species.

CHAPTER 1

Get Out of That

Ts'ui Pên must have said once: *I am withdrawing to write a book*. And another time: *I am withdrawing to construct a labyrinth*. Every one imagined two works; to no one did it occur that the book and the maze were one and the same thing.

— Jorge Luis Borges

THE MOST FAMOUS maze of all probably never existed and, even if it had, would have been a doddle to solve – if depictions of it on Cretan coins are anything to go by. The architect Daedalus, the story goes, built a winding series of passageways called the Labyrinth for King Minos of Crete as a place to contain the Minotaur. This monstrosity, with a bovine head, a human body, and an understandably bad temper, was the offspring of Minos's wife and a white bull given to the king by Poseidon, god of the sea. As a punishment to the Athenians, whom he had defeated in battle, Minos demanded that a number of their young men and women be periodically sacrificed to the creature lurking at the Labyrinth's heart. One year, the hero Theseus of Athens took the place of one of the youths to be sacrificed, entered the dreaded system of chambers, unravelled a ball of thread given to him by Ariadne, Minos's daughter, as he went, slew the Minotaur, and then escaped by following the thread trail back to the entrance.

We don't know how the Labyrinth of Minos was laid out. In any case, it's just a legend – more bull than the actual product of human ingenuity. What we do have are coins from the island of Crete dating back to between 300 and 100 BCE that bear designs presumed to represent the layout of the famous lair of the Minotaur. Most of these depict a rather simple yet ingenious pattern, typically in the form of a seven- or eight-level unicursal maze. The number of levels is how many times you cross the path to the centre if you draw a line from the outside to the ultimate goal. 'Unicursal' means that there's only one way in and out. As for the distinction between 'maze' and 'labyrinth', that's a matter of what definition you choose.

Some languages have only one word for maze or labyrinth; the Spanish *laberinto*, for instance, translates as either. 'Maze' is Old English for 'confuse' or 'confound', while 'labyrinth' comes from the Greek *labyrinthos*, the etymology of which is controversial. Some scholars have linked the Greek term with an old Lydian word, *labrys*, for 'double-edged axe', a symbol of royal power. So, the theory goes, the Labyrinth was part of the palace of the double axe – the home of the Minoan kings. It's a tentative and questionable link. In any event, we're left with a choice of definition and how, if at all, we want to distinguish between a labyrinth and a maze.

Our purposes being mainly mathematical, we'll assume that a labyrinth is a special type of maze – a unicursal maze. A labyrinth, then, is just a winding passage with no choices of which way to go or leave (except back the way we came).

A maze, on the other hand, we'll take to be the general case of a system of paths that may have multiple branches and a layout as confusing and convoluted as the maze-designer cares to imagine. A maze may also have multiple entrances, exits, and dead ends, whereas in the form of a labyrinth, although it may be ingeniously long to traverse given its total area, it consists of nothing more than an unbranching path, which leads to the centre and then back out the same way, with only one point of entry and exit.

Labyrinths aren't so much an intellectual challenge as they are a place to spend time in an unusual environment. As such they tend to have been employed as a form of meditation, a point nicely captured in the phrase 'you enter a maze to lose yourself and a labyrinth to find yourself.' Not surprisingly, designs for labyrinths are found in places of spiritual reflection. A well-known one is set into the floor of the nave of Chartres Cathedral, the border made of blue-black marble and the path itself of 276 slabs of white limestone. With a diameter of just under 13 metres (about 42 feet) it's large enough for a person to walk around the snaking track, as pilgrims have done since its construction sometime in the early thirteenth century. Rumour has it that there was once a depiction of the Minotaur at the centre of the eleven concentric rings of the pattern, but the primary symbolism, for obvious reasons, is Christian. The design features four arms standing for the branches of the cross and a winding path intended to symbolise the road to Jerusalem. Those not able, or willing, to make an actual trek to the Holy City could thus simulate the journey, more conveniently, by walking around this handy representation or, for the genuinely pious, shuffling around it on their knees. Although not the most ornate or embellished of labyrinths found in ecclesiastical buildings around the world, the one at Chartres is considered archetypal and others like it are known as 'Chartres mazes'.

Many other labyrinthine patterns are to be found in other parts of the world from all periods of history, from the Neolithic and Bronze Ages to recent times. As we've seen, they're intended, not as puzzles to solve, but as devices for religious or spiritual practice, ritual, or ceremony. It's thought that, long ago, Nordic fishermen would walk labyrinths before heading out to sea as a way of ensuring a plentiful haul and safe return, and that, in Germany, young men did the same as a rite of passage to adulthood. But these motives for their creation and design don't detract from the mathematical interest of labyrinths. The ingenuity and variety of techniques used to pack such a long path into a comparatively small space are fascinating in themselves. There's the study,

A Chartres-style labyrinth in the Abbey of Our Lady of Saint-Remy, Rochefort, Wallonia, Belgium.

too, of all the different ways of producing unicursal paths from 'seeds' – the starting shapes in the form of short sections of curves in a symmetrical pattern, that determine the initial course of the path of the labyrinth, from the centre working out. Labyrinths can be left- or right-handed depending on the direction of the first turn after entering, have different numbers of circuits, and take any of a couple of dozen or more distinct forms (largely determined by the choice of seeds) known to specialists in the subject.

The first mathematician to do a thorough analysis of unicursal mazes was the prolific Swiss theoretician Leonhard Euler (pronounced 'oiler') in the mid-eighteenth century. His interest in the subject stemmed from an answer he presented to the St Petersburg Academy in 1736 to the problem of the Bridges of Königsberg. The question was whether it was possible to walk from anywhere in the East Prussian city of Königsberg (present-day Kaliningrad in Russia) and cross every bridge there exactly once before returning to the starting point. Six of the bridges connected the banks of the river (three on either side) with two islands in the middle, while a seventh joined the islands. Euler reduced the problem to its mathematical essentials and, in this way, made it much easier to solve. He realised that the only information of relevance had to do with the connections: each land mass could be thought of as a point and each bridge a line joining two points. Euler was able to prove that for *any* arrangement of points and connecting lines, it would be possible to arrive back at the starting point, having traversed every connecting line exactly once, if and only if a certain condition was satisfied. This condition was that either no point along the way had an odd number of connecting lines or only two points did. Since the Königsberg layout of bridges broke the rule, there was no way to solve the original problem of crossing all the bridges just once and returning to the place you began.

The beauty of Euler's approach to the famous problem was that it could be generalised. His Königsberg analysis provided the first clear mathematical definition of a unicursal figure, as one that obeys the rule of connectivity just mentioned. But, more importantly, his work on this puzzle gave birth to a whole new field of maths known as graph theory and was important, too, in the rise of another major nascent subject – topology.

Both graph theory and topology are among the tools that mathematicians can bring to bear when tackling the thornier issue of multicursal mazes. Not only are such mazes designed to pose a mental challenge but also they can be fiendishly hard to solve, exist in two, three, or more dimensions, and take a form that doesn't, at first sight, even look like a maze.

Outside of legend, the first maze of which there's any historical record is that referred to by the Greek historian Herodotus who lived in the fifth century BCE. He describes a maze in Egypt so grand that 'all the works and buildings of the Greeks put together would certainly be inferior to this labyrinth as regards labour and expense.' Whether it was actually a labyrinth in the sense of being unicursal, we don't know. But, if Herodotus is to be believed, it would certainly have been impressive, with 12 courts, 3,000 chambers, and one side consisting of a pyramid 243 feet high.

Among more recent puzzle mazes are those that European royalty had built on their properties to amuse guests or provide places for secret meetings and trysts. The one at Hampton Court Palace, on the banks of the Thames, commissioned in the 1690s, is the best known and has now become a popular tourist attraction. The oldest surviving hedge maze in Britain, with walls tall enough to block any view of the way ahead, it covers a third of an acre but isn't hard to solve. Though not unicursal, it has only a few places where the path forks so that no one can get lost for long. Daniel Defoe mentions it in *From London to Land's End*, as does Jerome K. Jerome in *Three Men in a Boat*:

> We'll just go in here, so that you can say you've been, but it's very simple. It's absurd to call it a maze. You keep on taking the first turning to the right. We'll just walk round for ten minutes, and then go and get some lunch.

Far more convoluted is Il Labirinto Stra (the labyrinth of Stra). Located just outside the city limits of Venice, in the grounds of the Villa Pisani, and created in 1720, it's reputed to be one of the hardest public mazes in the world to solve.

The octagonal Jubilee Maze at Symonds Yat, Herefordshire.

Even Napoleon, a smart guy and no mean mathematician, is said to have been baffled by it. Anyone, however, who manages to navigate their way through the nine concentric rings of the maze, with their multiple openings and branches, can then climb the spiral staircase of the turret at the centre to get a bird's-eye view of the whole affair.

Two record-breaking mazes are in the United States. The Dole Plantation's giant Pineapple Garden Maze in Hawaii, made up of 14,000 tropical plants bordering two and a half miles of paths, was declared the world's longest in 2008. Meanwhile, not to be outdone, Cool Patch Pumpkins in Dixon, California, grew a corn maze that earned an entry in the *Guinness Book of Records* as the largest such temporary maze. So confounded by it were some visitors that, fearing they wouldn't escape before closing time, they dialled 911 to be rescued!

Let's suppose, then, that you've just entered a maze, about which you know nothing, for the first time. You've no idea how big or complicated it is, the walls are too high to see over, and there's no one else around with whom to compare notes. All you've been told is that there's a goal – a place in the middle that you need to reach in order to solve the puzzle – and definitely at least one route that leads there. A classic and straightforward approach is the 'wall following' method, in which you keep your hand in contact with one side of the maze and just keep walking. This will work in many cases, in the sense that it'll eventually lead you to your goal. But it has two drawbacks. First, it may take you a very long time, and second, it may fail completely if the maze has loops in it as well as dead ends that aren't connected to the outer wall. The key to solving mazes in a systematic way that won't let you down is to turn to maths.

Following Euler's example, the first step towards successfully traversing a maze is to transform it into an abstract plan. We can do this by using ideas from a subject known as network topology. In negotiating a maze, all that matters is what we do at points where there's a choice available – the so-called decision points. The first decision point comes at the entrance because we can choose to go in or not! Dead ends are decision points, too, although they offer only the option of stopping or turning around. More interesting are where the path splits and we can opt to go down one of two or more branches. If a maze is shown as a network, in other words as a series of points connected by lines, it's easy to see the solution – the best way to get from the entrance to the centre. Complicated subway systems, such as the London Underground, are maze-like and confusing to those unfamiliar with them, but maps in the form of network diagrams, on station walls and in every carriage, make it clear how to travel from any station to the desired destination.

We're supposing, though, that you've entered a maze without the benefit of such a map. This is where a bag of popcorn and a bag of peanuts come in handy – and not as snacks to be eaten if you end up getting lost! The popcorn and peanuts are to lay down trails so that you can take advantage of what Euler discovered from his work on the Königsberg problem. The trick is to ensure that, whatever choices you make at decision points, you never go along any stretch of path more than twice. So here's the method: leave a trail of popcorn as you go and a piece of popcorn at every decision point. That way you'll know if you've been down that path and to that point before. If you choose to go down a path a second time, then leave a trail of peanuts. The rule is, if you come to a path that you've already marked with peanuts, don't go down it again. Now, for a bit of nomenclature. If you arrive at a decision point that's popcorn-less, call this a *new node* and put down a piece of popcorn, thereby turning it into an *old node*. In the same way, if you come to a path that has no popcorn on it this is a *new path*. If you walk down it, drop popcorn as you go. The next time you walk down it, trail peanuts so that it becomes an *old path*.

With all this in mind, here's how you crack the maze. At the entrance, choose any path. When you come to a new node go down any new path. If you're on a new path and you come to an old node or a dead end, turn around and go back along the path. If you're on an old path and you come to an old node, take a new path if there is one or an old path otherwise. Don't go down a path twice. Follow these steps, make sure you're well stocked with popcorn and peanuts, and you're bound to reach the centre. You can then turn around and follow the route that's marked only by popcorn to find your way back out again.

A series of well-defined instructions that gives a guaranteed solution to a certain class of problems is known as an algorithm. This one for solving mazes is called Trémaux's algorithm, after the nineteenth-century French author Charles Trémaux who first described it. It's now recognised as a version of what's called depth-first search (DFS) – a method that can be used to search data structures known in maths as trees or graphs. Both of these structures consist of points, or nodes, which are linked by connecting lines, or 'edges'. Graph theory, in particular, which, as we mentioned, sprang from Euler's work on the Königsberg problem, is the source of a number of algorithms useful in tackling mazes. It's also a powerful tool for representing as mazes problems that don't superficially look like mazes at all – like Rubik's Cube.

Astonishingly, a standard $3 \times 3 \times 3$ Rubik's Cube has 43,252,003,274,489,856,000 possible arrangements. Each of these positions corresponds to a decision point in a maze of fiendish complexity. Simply spinning a cube at random would be as likely to succeed as a drunk staggering around a planet-sized maze in the hope of reaching the centre. The key to solving the puzzle in a reasonable amount of time lies in applying algorithms so that more pieces are brought into position without disturbing those already in place.

In graph theory there's a concept known as graph diameter. This is the largest possible number of nodes that have to be passed through in order to travel from one node to another when paths that backtrack, detour, or loop around are ignored. In the case of Rubik's Cube this equates to the maximum number of moves needed to solve the puzzle from *any* starting position (including the most randomised, worst-case scenario). Although the Cube was invented in 1974 it took until 2010 for its graph diameter, sometimes referred to as God's Number, to be calculated. Eventually, a team of researchers at Google, having burned through 35 CPU-years of computer time, found the answer: just 20. This surprisingly low number explains how top 'speedcubers' can solve the puzzle in under five seconds (the world record is 4.22 seconds, from a random starting position, set by a 22-year-old Australian in 2018). At least, it explains how it's *physically* possible. The real key to such extraordinary proficiency is endless hours of practice and memorising the steps involved in various efficient algorithmic strategies. To these demands must be added exceptional memory in the case of cubers who are able to solve the puzzle blindfold.

Complex mazes sometimes arise naturally on Earth, providing plenty of opportunity for people to get lost. In

South Florida, large stands of mangrove forming impenetrably thick walls and rising to a height of 70 feet line the sides of twisting channels. Although the waterways may not be long, a kayaker who enters without a guide or a map is liable to end up going round in circles for hours. Geological formations too can form natural mazes, which often become popular tourist attractions. Rock Maze, near Rapid City, in the Black Hills of South Dakota, includes an area of massive granite boulders that have separated and cracked to create a network of narrow, sinuous passageways.

When mazes form underground, as twisting interconnected systems of caves, they can have the added complication of being three-dimensional. Among the most extraordinary examples is the Optymistychna Cave near the Ukrainian village of Korolivka. Discovered as recently as 1966 the cave is confined to a layer of gypsum less than 30 metres (98 feet) thick and consists mostly of small passages that are no more than 3 metres (10 feet) wide and 1.5 metres (5 feet) high, although they can rise taller at intersections. To date, more than 265 kilometres (165 miles) have been mapped, making it the fifth longest known cave in the world. Longest of all – by a wide margin – is Mammoth Cave in central Kentucky, with passageways stretching for 663 kilometres (412 miles) through limestone that dates back more than 300 million years.

One of the amateur cavers who, in the early 1970s, helped create survey maps of Mammoth Cave, was Will Crowther, a programmer with the R&D company Bolt, Beranek and Newman. Crowther was part of the original, small team that developed ARPANET (a forerunner of the Internet). A fan of the tabletop role-playing game Dungeons & Dragons, he hit upon the idea of combining a computer simulation of his

Rotunda Room, Mammoth Cave, Kentucky. USGS photo.

caving explorations with elements of fantasy role-playing. The result, developed in 1975 and 1976, was *Colossal Cave Adventure*, which became popularly known as *Adventure* or simply 'Advent' (after the name of its executable file). Crowther's original 700 lines of FORTRAN code were expanded by Don Woods, a graduate student at Stanford University, who added more fantasy ideas and settings based on his love of Tolkien's writings. By 1977, the canon version of *Adventure* was complete and soon became widely distributed among programmers in the US and elsewhere. Its 3,000 lines of code were supplemented by 1,800 lines of data, which included 140 map locations, 293 vocabulary words, 53 objects (15 of them treasure objects), travel tables, and various messages, the most famous of which has become: 'You are in a maze of twisty little passages, all alike.' Part of the fun of the game was in figuring out how to map this maze with pen and paper. A useful approach was to drop objects in the rooms you encountered as you went along to serve as landmarks.

Our look at cave mazes wouldn't be complete without mention of the Labyrinthos Caves, located beneath a quarry at Gortyn in the south of Crete, just 20 miles or so from the Minoan palace at Knossos. This series of chambers and tunnels, some researchers claim, may be the true source of the legend of the Minotaur. Visitors can explore up to two and a half miles of intertwining passageways, which occasionally open into large rooms such as the Altar Chamber. Whether this natural maze inspired the famous legend we'll probably never know, but the Labyrinthos Caves aren't short of fascinating historical episodes as it is, including a time when the spies of Louis XVI carried out covert operations from there and the Nazis employed them as a secret ammunitions dump in World War II.

Psychologists use mazes for experiments in animal cognition, while researchers in artificial intelligence challenge their robotic inventions to navigate mazes in the most efficient way. The maze that is the Internet is one of the most elaborate creations of the human mind, which, in turn, inhabits the maze of neurons and their interconnections, which make up our brain. Curiously, James Knierim of Johns Hopkins University and his colleagues found that, in some situations, such as when we try to recall if we've seen a person's face before, our brains work in a way that's similar to how a rat navigates its way out of a maze. Different areas of the hippocampus arrive at two different conclusions – either the face is familiar or it isn't – which are then voted on by other parts of the brain in order to come to a decision. The researchers found that a similar decision-making process takes place in a rat's brain if the animal is taught how to recognise a certain maze but then, later, some landmarks in the maze are slightly altered.

In creating mazes as challenges for the mind, or labyrinths for the purpose of contemplation, we're in a sense externalising the nature of our own brains and how they operate. The Argentinian writer Jorge Luis Borges repeatedly used the labyrinth as a metaphor for some of the great mysteries of the world, including time, mind, and physical reality. The epigraph for this chapter comes from his short story 'The Garden of Forking Paths' (1941), while in 'Ibn-Hakam Al-Bokhari, Murdered in His Labyrinth' (1951) one of the characters, Unwin the mathematician, remarks: 'There's no need to build a labyrinth when the entire universe is one.'

CHAPTER 2

At the Vanishing Point

I love to talk about nothing. It's the only thing I know anything about.

— Oscar Wilde

ZERO: THERE ISN'T much to it. Certainly not if it happens to be your bank balance, how many birthday cards you received, or a good approximation to your chances of winning the rollover lottery jackpot. At the same time, what zero is seems obvious. We all know what it means and take its existence for granted. It's hard to imagine that there was a time when mathematicians got by without it and that it actually had to be discovered – or invented, depending on how you look at it.

Intuitively, of course, the idea of zero goes back before the dawn of history. Early humans – even animals for that matter – know what it's like to have zero food or zero shelter. Having nothing, or the threat of having nothing, is what gives us the impetus to survive.

For philosophers, zero, and the concept of nothing, has long been a subject of fascination. The notion of the void plays an especially important role in many Eastern philosophies. In some forms of Buddhism, for instance, Śūnyatā (emptiness) is considered to be a state of mind in which all conscious thoughts, including awareness of self, are released leaving just pure consciousness of the moment. It's a state that, for example, practitioners of Zen archery strive for, so that the only focus left is the action of the shot.

According to other schools of thought, such as that of Aristotle, nothingness is an impossibility. After all, the argument goes, surely nothing is the one thing that *can't* exist. By its very nature it's the negation of existence (including that of space, time, matter, and energy): there always has to be something. From this position, Aristotle argued that the universe must be eternal since if it had been created at some point it would have to have been preceded by the one thing that he wouldn't countenance – a void. Other prominent Greeks disagreed.

Democritus and his followers believed that all matter was made of atoms. Therefore, they insisted, there had to be a void to provide the space in which atoms could move. Although, in time, scientists came to learn of the existence of atoms (albeit of a very different kind from those the classical atomists had in mind), it was Aristotle's philosophy that persisted and came to dominate mediaeval thought in Europe. During the Middle Ages, the Catholic Church was so afraid of the void and insistent on an eternal universe that it gave Aristotle precedence over the Book of Genesis. 'Nature abhors a vacuum' was the motto of this worldview, and it led early scientists to suppose that if ever a vacuum started to form it would exert a force pulling matter towards it in order to fill itself and thereby circumvent the dreaded void.

In mathematics, we're so used to the concept of nothing, or zero, from an early age, that it seems odd that it took so long for it to first emerge in the historical record. The truth is, though, that maths started out as a purely practical affair, as a way of keeping track of how many items you had, or owed, or were owed, or figuring out the size of things. I may need a way of reckoning that I have 18 horses or 43 sheep, and how many I'll have if I buy or sell several more of each. But why should I need to keep track of having none of something or figure out how many bricks I'll need to build a wall that has no height? Mathematical problems started out firmly rooted in real, everyday situations not abstract ones. It was the tool of the merchant, the government bookkeeper, and the architect, so that numbers used to have a much more concrete meaning than they do today. It's easy to overlook the huge mental leaps that are involved in going from eight specific things, such as eight amphorae of olive oil, to eight things in general, to eight as an intangible entity in its own right. There's no immediate obvious use for having the means to deal with zero things.

Our distant ancestors started out with the positive integers, 1, 2, 3 … Zero came much later and its evolution is uncertain and convoluted. How and when zero arose is also complicated by it having two different uses: as an empty place indicator and as a number with equal status to any other. In the number 3075, zero serves to put 3 in the correct position so that it means 'three thousand' rather than 'three hundred'. It takes on a completely different role if we allow it to exist as the number that is one less than one. Then it needs to be incorporated into our arithmetic as a number with certain properties. What happens when you add zero, multiply by it, or, most intriguingly, divide by it? Then there's the whole business of how to represent zero in these two quite distinct contexts – the notation to be used, and the name, and whether these ways of representing this new intellectual creation should be distinct depending on whether we're referring to the place-holding role or the number-in-its-own-right role. Our name 'zero', by the way, comes from the Arabic *sifr*, which is also the root of the word 'cipher'.

It was as a place-holder, to make clear the value of a multi-digit number, that the idea of zero first appeared in maths. Place-value number systems, in which the position of a digit indicates its value, go back at least 4,000 years to when the Babylonians started using them. But there's no evidence that these people also felt the need to have an empty place indicator, at least not for a very long time. Original texts from around 1700 BCE survive, in the form of cuneiform writing pressed into clay tablets with a stylus

Clay tablets, such as this one from southern Iraq, dating to 3100–3000 BCE, contain the earliest known symbols for zero.

to leave wedge-shaped marks. These tablets reveal how the Babylonians represented numbers and did arithmetic with them. Their notation was quite different from ours and their number system was based on 60 rather than 10. But it's clear that the early Babylonians didn't distinguish between what we would write as, say, 1036 and 136, except by context. It wasn't until around 700 BCE that they started to include symbols in the same way that we do for the idea of zero as a place-marker. Various notations were used, according to the city and the era, but one, two, or three wedge-shaped symbols can be seen on Babylonian and Mesopotamian tablets in place of where we would put a '0'. The same idea occurred later to other civilisations, including the Chinese, who left an empty space as the equivalent of zero in their counting-rod system, and the Mayans.

The drawbacks of not having a place-value number system quickly become obvious if we try doing maths using the alternative – where each symbol stands for a value that's fixed and can't be changed. The Romans were lumbered with such an approach, which is perhaps why we hear a lot about Roman generals, politicians, conquests, and methods of government and town planning, but not so much about Roman breakthroughs in mathematics. Roman numerals use seven letters as symbols for writing numbers: I for 1, V for 5, X for 10, L for 50, C for 100, D for 500, and M for 1,000. Their use gets cumbersome very quickly. For example, 1,999 in Roman numerals is MCMXCIX and any number much larger than 5,000 can be ridiculously hard to represent. The other big issue is doing arithmetic Roman style. For us, figuring out that $47 + 72 = 119$ is pretty straightforward. But try adding XLVII and LXXII. The easiest way is to convert the Roman numerals into our decimal (base-10) system and then convert back to give the answer, CXIX. Doing it the Roman way is tortuous, and as for multiplication ...

Zero as a number in itself, rather than as a place-holder, is a much more recent invention (or discovery). Its introduction in this guise is attributed to Pingala, an Indian scholar who lived during the third and second centuries BCE. Pingala used a place-value notation, based on binary rather than decimal because a binary notation allowed numbers to be encoded into Sanskrit verse. But he also employed the word 'sunya', Sanskrit for empty, to refer to the number zero. The earliest appearance of the symbol in its modern form is in the Bakhshali manuscript, a text written on birch bark and found in the summer of 1881 near the village of Bakhshali in what at the time was British-ruled India but is now Pakistan. A large part of the manuscript had been destroyed and only about 70 leaves of bark, of which a few were mere scraps, survived to the time of its discovery. From what we can gather it seems to be a commentary on an earlier mathematical work, setting out rules and techniques for solving problems, mostly in arithmetic and algebra, but also to a lesser extent in geometry and mensuration (the maths of measurements). Now kept at the Bodleian Library in Oxford, the manuscript has been recently carbon-dated to the third or fourth century CE making it several centuries older than previously thought.

Later, in the seventh century CE, the Indian mathematician Brahmagupta put the concept of zero as a number on a firm footing. He laid down various rules for doing arithmetic when zero and negative numbers (another innovation at the time) were involved. Most of his rules now look very familiar. For example, he argued that the sum of zero and a negative number is negative, the sum of a positive number and zero is positive, and the sum of zero and zero is zero.

Regarding subtraction, his rules are also those we still use today: a negative number subtracted from zero is positive, and so on. But with division, he ran into difficulties. Zero divided by zero he thought should be zero. The value of any other fraction, however, with either zero on top and a positive or negative number on the bottom, or vice versa, was a mystery to him.

Brahmagupta wouldn't commit to saying what happens if you divide, for example, eight by zero. This isn't surprising because it's by no means clear what the answer should be. Five hundred years later, another Indian mathematician (and astronomer), Bhaskara, claimed in his great work Siddhānta Shiromani ('Crown of Treatises') that the result of dividing a number by zero was 'an infinite quantity'. Thereafter he waxed lyrical about the philosophical justification:

> In this quantity consisting of that which has zero for
> its divisor, there is no alteration, though many may
> be inserted or extracted; as no change takes place
> in the infinite and immutable God when worlds are
> created or destroyed, though numerous orders of
> beings are absorbed or put forth.

It's certainly possible to see some logic behind Bhaskara's desire to put a number divided by zero equal to infinity. After all, if we divide any number, say, 1, by numbers that get smaller and smaller, the answer gets bigger and bigger. The trouble is if we let $n/0 = \infty$, where n is any finite number, then 0 times ∞ can equal any number at all, which doesn't make much sense. In fact, there are many little mathematical tricks that involve a concealed division by zero which make it seem that you can prove $1 = 2$, or, more generally, that any number is equal to any other number. To avoid this kind of confusion and inconsistency, mathematicians eventually decided that division by zero isn't allowed, or to be more precise, that the result is undefined.

In modern maths, there are plenty of concepts that are related to zero without actually being zero. One of these is the empty set. In set theory, the empty set is, unsurprisingly, the set that has no members. This is a different concept from zero itself, most obviously in the fact that the empty set *is* a set while zero is a number. But zero is the number of elements, or *cardinality*, of the empty set. Sets have their own operations that act in a similar way to addition and multiplication, and are known as union and intersection. The union of two sets is simply the set that contains all elements in at least one of the two sets, and the intersection is the set that contains all elements in *both* of the two sets. The empty set plays a role analogous to that of zero: the union of any set with the empty set is the set itself (just as $x + 0 = x$) and the intersection of any set with the empty set is the empty set (in the same way that $x \times 0 = 0$).

Other things related to zero turn up when we try to get as close as possible to zero without ever reaching it. One path that takes us on this quest is the sequence 1, 1/2, 1/4, 1/8 ... in which each term is half the size of the previous one. We'd normally say that the *limit* of this sequence – the value to which it's converging – is simply 0. But can there be such a concept as 'infinitely close' to zero without actually arriving there? The system of real numbers, which includes all the points on the number line, provides us with no such concept. The best it can do is supply numbers as small as we desire. Yet no matter how fantastically small it is, a non-zero real number will always be finitely small not infinitely small. To achieve the goal of being able to approach infinitely close to zero we need a new type of number, something that lies both beyond the scope of imagination and our conventional way of reckoning.

The English mathematician John Conway was looking for a fresh approach to help him analyse certain types of games. In a moment of inspiration, while watching the British Go champion play in the mathematics department at Cambridge, he saw a way to make progress. Conway noticed that endgames in Go tend to break up into a sum of games, and that some positions behaved like numbers. He then found that, in the case of infinite games, positions arose that behaved like a new kind of number, which became known as the surreals. The name was coined not by Conway himself but by the American mathematician and computer scientist Donald Knuth in his 1974 book *Surreal Numbers: How Two Ex-Students Turned on to Pure Mathematics and Found Total Happiness*. It's a novelette notable for being the only instance where a major mathematical idea was first made public in a work of fiction.

A surreal number is a member of a mind-bogglingly vast community of numbers. Included in this are all of the real numbers, a host of infinitely large numbers known as infinite ordinals, a set of infinitesimals (infinitely small numbers) produced from these ordinals, and strange numbers that previously lay outside the known realm of mathematics. Each real number, it turns out, is surrounded by a 'cloud' of surreals that lie closer to it than do any other real numbers. One of these surreal clouds inhabits the twilight zone between zero and the smallest real number greater than zero, and is made up of the infinitesimals. These infinitely small numbers have values less than any number in the sequence

1, 1/2, 1/4, 1/8 … no matter how far we go along it. One such surreal number is ε (epsilon), which we can define as the first surreal number to be greater than 0 but less than 1, 1/2, 1/4, 1/8…

In Knuth's fictional tale, two college graduates, Bill and Alice, are living on an island in the Indian Ocean to get away from civilisation when they come across a black rock, half buried in the sand, with writing on it. Bill begins to read: 'In the beginning everything was void, and J.H.W.H. Conway began to create numbers. Conway said, "Let there be two rules which bring forth all numbers large and small …"'

Day by day, Bill and Alice work through the stone's inscriptions and learn how to build a whole new system of numbers, inconceivably more vast than anything they'd previously imagined. The basic idea of this new system is that any real number N can be represented using two sets: a set L (left) which contains numbers less than N, and a set R (right) which contains numbers greater than N. (We'll look in more detail at the workings of this process in Chapter 6.) Using Conway's two rules, the stone explains how, starting with nothing at all, the number zero can be created from an empty left set and an empty right set. More numbers can then be brought into existence by putting zero in the left set of one number and the right set of another, and using these new numbers to create still more numbers. Eventually, all the members of this spectacularly huge collection of numbers – the surreals – are formed, among which are the infinitesimals.

Ultimately, asking what is the closest we can get to zero that's not zero is similar to asking what's the closest we can get to infinity that's not infinity. Using real numbers on their own, it doesn't make sense to talk about infinitesimals, because if you try to name any number, no matter how small, there'll always be a smaller number between it and zero. It's the same with big real numbers: there is no biggest one as there's no limit to how large you can get. There are real numbers between any number you can name and infinity. Fortunately, in our search for the infinitely small and the infinitely large, the mathematical universe is huge and diverse enough that it lets us conjure up new number systems that make the previously impossible a reality.

One surprising fact in maths, which at first doesn't look right, is that $0.999... = 1$. It seems at odds with common sense because 0.9, 0.99, and so on are all less than 1, so it would seem that $0.999...$ (where the nines go on forever) should also be less than 1. Yet there are many easy ways to show that $0.999... = 1$. For instance, let $x = 0.999...$. Then $10x = 9.999... = x + 9$. Subtracting x gives $9x = 9$, so that $x = 1$. We've proved in a few simple steps that $0.999... = 1$ and, at the same time, that $1 - 0.999...$ isn't some very tiny number, or even an infinitesimal, but instead is exactly equal to 0.

To grasp this strange result properly, we need to understand what's meant by $0.999...$, or, for that matter, any real number with an infinite decimal representation. By the decimal expansion, for example, of pi – $3.14159...$ – we mean the limit of the sequence 3, 3.1, 3.14, 3.141.... In the same way, we can define all real numbers using only those rationals with a terminating decimal sequence. (Not all rationals have such a terminating sequence; for example 1/3.) So, $0.999...$ is the limit of 0.9, 0.99, 0.999... and so on, which is indeed exactly equal to 1.

Surreal numbers shed a whole new light on this matter. In the case of the surreals, only certain numbers are defined in a finite number of steps. These are the so-called dyadic rationals, which are fractions whose denominator is a power of 2. For this reason, it makes more sense to use binary when working with such surreals. The equivalent of 0.999... in binary is 0.111... which is 1/2 + 1/4 + 1/8 and so on, which is again equal to 1. We know what infinite decimal (or binary) representations mean with real numbers but with surreal numbers it's a different matter. For example (using decimals for simplicity) π = 3.14159... How is this written in surreal numbers? It's certainly greater than 3, 3.1, 3.14, and so on, but then again so is 4, and in surreal numbers simply using these alone will just return the answer 4. Likewise, π is certainly less than 4, 3.2, 3.15, and so on but in surreals using these will just give 3. We need to use both together to home in on the precise value of π, which would be represented as {3, 3.1, 3.14, ... | 4, 3.2, 3.15, ... }.

So what does this mean for 0.999..., or rather 0.111... in binary? In surreal numbers (using binary), this would be {0.1, 0.11, 0.111, ... | 1.0, 1.00, 1.000, ... } The set L seems to approach 1 and indeed would have limit 1 in the real numbers but the set R only really contains one number, namely 1. Thus, we get the bizarre conclusion that this is actually less than 1, and it turns out to be exactly $1 - \varepsilon$. Then again, 1.000... turns out to be greater than 1, and is $1 + \varepsilon$. This also shows that when dealing with surreal numbers, representations in decimal or even binary are not the most helpful way to think about a number, and we should really be thinking about the sets L and R.

When calculus was developed by Isaac Newton and, independently, by Gottfried Leibniz, there was a problem that seemed to persist. This was how to describe changes that get smaller and smaller without invoking 0/0, which is undefined. Bishop George Berkeley, an early critic of Newton's calculus, remarked:

> And what are these fluxions? The velocities of eva-
> nescent increments? And what are these same eva-
> nescent increments? They are neither finite quanti-
> ties, nor quantities infinitely small, nor yet nothing.
> May we not call them ghosts of departed quantities?

Newton could describe the rate of change over intervals that were as small as necessary, and could clearly see that they approached, ever more closely, a specific value. The difficulty came in proving that this was indeed the true value without resorting to infinitesimals. Newton used the letter o to denote an arbitrarily small number added to a quantity x in order to find the rate of change. He then blotted out any terms containing o because they were negligible, yet the terms themselves had a non-zero, albeit arbitrarily small value. How could you simply cancel them out? That was the major criticism. Ultimately, while the rest of mathematics was built on firm logic, calculus rested on faith, as any attempt to make calculus rigorous inevitably either invoked 0/0 or resulted in having to simply ignore very small but non-zero terms, treating them as zero.

Nowadays, in calculus, to avoid dealing with infinitesi-mals, we use limits, a method developed by French mathe-matician and philosopher Jean le Rond d'Alembert in the mid-eighteenth century. A limit is an endpoint we head towards if we let a variable (usually denoted by x) approach some number closer and closer, without ever reaching it. It's a technique by which we can avoid the ultimate mathematical embarrassment – finding ourselves dividing by zero. Suppose we want to know what happens to $x^2 - 1$ divided by $x - 1$ as x approaches the value 1. We can't simply plug in $x = 1$ and jump to the result in one quick step, because we end up with 0/0. Instead, we have to let x creep up on the value 1, bit by bit: when $x = 0.5$, $(x^2 - 1)/(x - 1) = 1.5$; when $x = 0.9$, $(x^2 - 1)/(x - 1) = 1.9$; when $x = 0.999$, $(x^2 - 1)/(x - 1) = 1.999$, and so on. The endpoint is clearly 2, even though we can't, in this case, put in the value for x that transports us instantly to this final answer. It's simply the limit of the process.

In some ways, approaching zero closer and closer in maths is analogous to the efforts of physicists to produce an ever more perfect vacuum – a space devoid of all matter. Those efforts began in earnest when seventeenth-century Italian physicist and mathematician Evangelista Torricelli learned of the fact that no matter how strong a team of workers were, they couldn't use their water pumps to draw water more than 10 metres vertically. In 1643, Torricelli decided to try this experiment using mercury instead of water, as mercury, being denser, would involve a much smaller height. He found in this

Torricelli experimenting with a barometer in the Alps, 1643. Oil painting by Ernest Board.

case the limit to be around 76 centimetres. He then took a tube that was slightly longer than 76 centimetres, sealed one end, filled it with mercury, and placed the upturned tube in a bowl that also contained mercury. Whenever he did this, the mercury level in the column would always drop to 76 centimetres. As air couldn't enter the space at the top of the column – the open end of the column being submerged in mercury – Torricelli deduced that he'd created a vacuum. To be sure, it wasn't a perfect vacuum (for one thing, it would contain traces of mercury vapour) but it was good enough to give the lie to the ancient philosophical claim that Nature abhorred a vacuum.

The total height of the column didn't affect the height of the mercury but when Torricelli tried the same experiment part way up a mountain, he noticed that the mercury level was lower. Torricelli realised that it was the air that was pushing as opposed to the vacuum pulling, from which he concluded: 'We live submerged at the bottom of an ocean of air.' His discovery came as one of the final blows to Aristotle's (and the mediaeval Church's) worldview. In response to the claim that vacuums couldn't exist, Torricelli had simply gone ahead and created one.

But times move on. In classical physics, the only kind that Torricelli, Newton, and every scientist before the turn of the twentieth century knew about, perfect vacuums are theoretically possible. We may lack the technology to remove every last air molecule from a sealed container, but it's at least conceivable that we could do so. The result would be a volume of space devoid of even a single particle of matter. With the dawn of quantum mechanics however (the subject of Chapter 9) all preconceived notions of space and time, matter and energy, were shattered. In this startling new vision of physics, the possibility of ever achieving a true void – a place where there's absolutely no material particles or energy of any kind – was ruled out, once and for all.

A so-called *quantum vacuum*, which is the ultimate nature of the space in which we live, is seething with particles. These aren't the bits of matter – electrons, protons, neutrons, atoms, ions, and molecules – that make up the conventional physical universe we see, but instead are 'virtual particles'. These ephemera pop in and out of existence spontaneously, provided they're not being observed, before disappearing again, leaving no trace. The reality of virtual particles is guaranteed by Heisenberg's uncertainty principle, a tenet of quantum mechanics, which insists that it's not possible to know the position and momentum of a particle exactly: the more accurately you measure a particle's position, the less information you can have about its momentum. The same idea applies to the pairing of energy and time: the more accurate your measurement of energy, the less accurate will be your measurement of time. As a result of Heisenberg's principle, there'll always be an uncertainty in the measurement of energy (which is equivalent to mass by the famous equation $E = mc^2$), so that particles can momentarily materialise and then disappear before we have a chance to observe them. The quantum vacuum seethes with the comings and goings of virtual particles, making it impossible ever to achieve the perfect emptiness of a classical void.

Did just such a quantum fluctuation – the spontaneous appearance of a particle out of nothingness – spark off the whole universe? Cosmologists nowadays invoke such an idea as one explanation of how everything we see around us came about in the first instance. First, there was nothing. Then, the next moment, a quantum jitter set the whole cosmos into motion. It's an interesting thought: a modern take on the old conundrum of creation *ex nihilo*. But much remains unexplained. Before this universe we inhabit sprang into being, there must have been something. Nothing – zero in a physical sense – can't exist. Even if there was nothing of substance, there must at least have been the laws of quantum physics and, ultimately, the laws of mathematics behind these, which caused nothing to become something.

CHAPTER 3

Seven Numbers That Rule the Universe

You have to be odd to be number one.

– Dr Seuss

SEVEN? WHY NOT a nice 'round' number like 10? The only reason we think 10 *is* a round number and special is that we have ten fingers so that we've developed our most commonly used number system around it. If humans had eight fingers to count on, our maths would almost certainly be based on octal rather than decimal. So, if we're being impartial, seven is as good as any for the size of our elite group of cosmic conquering numbers.

According to Sheldon Cooper, *The Big Bang Theory*'s geek-in-chief, the best number is 73. Why?

> Sheldon: 73 is the 21st prime number. Its mirror, 37, is the 12th and *its* mirror, 21, is the product of multiplying 7 and 3.
> Leonard: We get it, 73 is the Chuck Norris of numbers!
> Sheldon: Chuck Norris wishes. In binary 73 is a palindrome, 1001001, which backwards is 1001001. All Chuck Norris backwards gets you is Sirron Kcuhc!

While Sheldon often sports a shirt with '73' on the front, fans of Douglas Adams's *The Hitchhiker's Guide to the Galaxy* might prefer '42' on theirs. It is, after all, the answer to life, the universe, and everything, arrived at by the megacomputer Deep Thought after seven and a half million years of cogitation. Those wanting to justify the choice of 42 might point to the fact that the element molybdenum, with atomic number 42, is coincidentally the 42nd most common element in the cosmos, or that the three best-selling albums in the world – Michael Jackson's *Thriller*, AC/DC's *Back in Black*, and Pink Floyd's *The Dark Side of the Moon* – all run for 42 minutes. In truth, as Adams himself explained, it was just a bit of fun: 'It had to be a number, an ordinary, smallish number, and I chose that one. I sat on my desk, stared into the garden and thought 42 will do. I typed it out. End of story.'

Joking aside, what *are* the top numbers in the universe? That depends, of course, on what we mean: the most commonly occurring, the most interesting (for whatever reason), the most important in maths? There's really no such thing as an uninteresting number. On one occasion the British mathematician G. H. Hardy was riding in a taxicab numbered 1729 on his way to visit the Indian prodigy Srinivasa Ramanujan (of whom more in Chapter 8), lying ill in a London hospital. After greeting Ramanujan, he remarked what a dull number 1729 seemed to be. Without hesitation, Ramanujan corrected him: 'It is a very interesting number. It is the smallest number expressible as the sum of two cubes in two different ways.' ($1729 = 1^3 + 12^3 = 9^3 + 10^3$.) Logic also dictates that completely uninteresting numbers can't exist. If they could, then there'd be a smallest uninteresting number, which would immediately become interesting due to its record smallness! It would then be replaced by a new smallest uninteresting number, which would again flip to being interesting for the same reason, and so on.

Physics has some important numbers that, at first glance, you might think would justify inclusion in our list – the speed of light, the universal gravitational constant, and Avogadro's Number among them. But most of these depend on the system of units that are used. In some ways, the speed of light in a vacuum is the most important quantity in physics. But its numerical value differs according to whether it's in kilometres per second (299,792), miles per second (186,282), or some other unit. The only constants in physics with values that don't depend on units are so-called dimensionless constants. One of the most important of these is the fine-structure constant, α, equal to almost exactly 1/137, which has a habit of popping up all over the place in atomic and subatomic physics. One way to think of it is as a measure of the strength of the electromagnetic interaction between elementary charged particles such as electrons, but it has many interpretations and appears to have a deep significance to the universe in which we're embedded – a significance that has yet to be properly fathomed. It's fascinating not just because of its ubiquity but also because (along with a couple of other factors) it involves a combination of three fundamental constants of nature: the square of the charge of the electron divided by the product of Planck's constant and the speed of light. In other circumstances, the fine-structure constant might make our list of top seven numbers in the universe. But this being a book mainly about maths, rather than physics, α must settle for an honourable mention.

Two numbers pretty much select themselves for membership in our elite band of numerical celebrities because of their fame and tendency to appear everywhere you look in maths. They're The Beatles and The Rolling Stones of the number scene: pi and e. Pi is the one more familiar to non-mathematicians because we all encounter it in school. It's the ratio of a circle's circumference (C) to its diameter (d), $\pi = C/d$. This in itself seems surprising at first: *why* is the ratio always the same regardless of the size of the circle? The answer is that all circles (on a flat plane, at least) are similar, a mathematical term meaning that they're all scaled versions of each other. The formula for the area (A) of a circle, $A = \pi r^2$, where r is the radius, also involves π as can be shown by cutting the circle into smaller and smaller pieces, then arranging the pieces to approximate a shape of which we can easily calculate the area.

We expect pi to turn up whenever circles are involved because its geometric roots lie in this shape. But the great wonder of pi is its habit of materialising, as if by magic, even when there's no circle in sight. For instance, the series $1/1^2 + 1/2^2 + 1/3^2 + 1/4^2 + 1/5^2 \ldots = 1 + 1/4 + 1/9 + 1/16 + 1/25 \ldots$ gets closer and closer to the value $\pi^2/6 = 1.645\ldots$, as we include more and more terms. Turn this fraction upside-down and we get $6/\pi^2$, which is equal to the probability that two numbers, providing they're big enough, are coprime – in other words that they have no common factors other than 1. Pi, in fact, is intimately and, somewhat mysteriously, involved with how prime numbers (numbers that have no factors other than themselves and 1) are distributed. It somehow ends up in a formula featuring one of the most important objects in maths, known as the Riemann zeta function, about which we'll have more to say in Chapter 13. Why should a number that, in the first instance, we meet as a basic property of circles suddenly re-emerge in connection with prime numbers?

Pi also crops up in answer to a problem known as Buffon's needle, a question first posed in the eighteenth century by French naturalist Georges-Louis Leclerc, who later became Count Buffon. Suppose a floor is made from parallel strips of wood that are the same distance l apart. If a needle, also of length l, is dropped onto this floor, what's the probability that it will land so that it crosses one of the lines between the wooden strips? The answer turns out to be $2/\pi$.

Back in 1655, English clergyman and mathematician John Wallis (who introduced the symbol ∞ for infinity) found that

$$\pi = 2 \left[\frac{2}{1} \cdot \frac{2}{3} \cdot \frac{4}{3} \cdot \frac{4}{5} \cdot \frac{6}{5} \cdot \frac{6}{7} \cdot \frac{8}{7} \cdot \frac{8}{9} \cdots \right]$$

Fast forward to 2015 and two researchers at the University of Rochester, Carl Hagen and Tamar Friedmann, were astonished to discover the very same formula emerge from calculations to do with the energy levels of a hydrogen atom. Hagen, a particle physicist, had been teaching a technique in quantum mechanics known as the variational approach to a class of students. It's a method that can be used to approximate the energy states of electrons in complicated systems, such as molecules, where exact solutions are impossible. Hagen thought it would be a good exercise for his students to apply the variational approach to the relatively simple case of a hydrogen atom, for which the energy levels *can* be calculated exactly, to see what errors were involved in using the approximation. When he looked at the problem himself, he spotted a pattern almost right away. The error in using the variational approach was 15 percent for the lowest energy state of hydrogen, 10 percent for the next lowest state, and kept on decreasing for each successive state. Hagen asked his colleague Friedmann, a mathematician, to look at how the approximation trended for higher energy levels. The limit approached by the technique as the energy levels increased exactly matched the Wallis formula.

Pi is no stranger to physicists. It's found in Coulomb's law of electric charge, Kepler's third law of planetary motion, and the field equations in Einstein's general theory of relativity, to name a few. Whenever circles, spheres, or periodic motion, which stems from circular motion, enter the picture, so too does pi. But it makes an unexpected entrance even when there are no circles or sine waves to be seen, as in the case just mentioned or Heisenberg's uncertainty principle. Sometimes a connection with the circular origins of pi can eventually be figured out but at other times there's no obvious link with the geometry of our schooldays: pi is just an omnipresent fact in both the material and mathematical universe.

The same is true of another, in some ways similar but less familiar number, which makes it into our list of the top seven. e, also known as Euler's number, has the value 2.71828…, a tad less than that of pi, and like pi is both irrational and transcendental. Irrational means that it can't be written in the form of one whole number divided by another. Transcendental means that it isn't the solution of an equation such as $x^3 + 4x^2 + x - 6 = 0$, in other words a polynomial with whole-number (or rational) coefficients.

Unlike in the case of pi, there's no one obvious definition of e. It comes out of many formulae, any one of which could serve as a definition. But a simple way to understand it is to think about a problem in compound interest. In fact, e was first encountered in this very way by the Swiss mathematician Jacob Bernoulli, in 1663. Suppose you deposited £100 in a bank which offers a 100% annual interest rate and pays annually. At the end of the year, you'd have £200. Now suppose instead you go to another bank which offers the same generous rate but pays biannually. You'd receive 50% compound interest every six months, and at the end of the year would have £225. Clearly, there's an advantage in having the compound interest paid more frequently. If the interest accrues monthly, you'd have £261.30 by year's end, and if paid daily, you'd have £271.46. Compounding the interest at shorter and shorter intervals helps, but there's a limit to the advantage you can gain. In fact, if the interest were compounded continuously, by the end of the year you'd have 100 times e, or £271.82, rounded to the nearest penny.

Another situation where e turns up has to do with exponential growth. An exponential curve is one defined by a number raised to the power x. The slope, or steepness, of such a curve increases as x gets bigger. The slope of the exponential curve 2^x, at any point x, is approximately 0.693×2^x; the slope of 3^x is approximately 1.098×3^x. Always, with

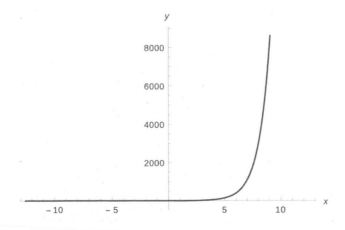

The graph of $y = e^x$

exponential curves, the slope is proportional to the height. But there's one special case where the slope is exactly equal to the height and this is the curve e^x. Not only is the steepness of the curve e^x equal to its height at every point, but so is the rate of increase of the steepness, and the rate of increase of the rate of increase of the steepness, and so on.

Like pi, e has a habit of appearing when least expected and in areas of maths that seem totally unconnected. Suppose you have two packs of playing cards. You shuffle both of them independently, then deal out the first card of each deck, the second card, and so on. What's the probability that there's no match – that no two consecutive cards dealt are the same? The answer is almost exactly equal to $1/e$. In fact, it's given by $1 - 1/1! + 1/2! - 1/3! + 1/4! - \ldots - 1/51! + 1/52!$ (where '!' means factorial; for example $4! = 4 \times 3 \times 2 \times 1$), which is within $1/53!$ of $1/e$. The probability of no match gets closer and closer to $1/e$ as the number of cards in the two decks is increased, providing there's only one of each card in each deck.

Google, the firm behind the ubiquitous search engine, has a particular fondness for e. In its stock market launch in 2004 its stated goal from its initial public offering was to raise e billion dollars, or (to the nearest dollar) \$2,718,281,828. Later, in a quest for talented recruits, it put up billboards in Silicon Valley, Seattle, Austin, and Cambridge, Massachusetts, that read '{first 10-digit prime found in consecutive digits of e}.com'. Anyone mathematically savvy enough to figure out the cryptically named website could visit it to find this message:

> Congratulations. You have made it to level 2. Go to www.linux.org and enter *Bobsyouruncle* as the login and the answer to this equation as the password.

$F(1) = 7182818284$
$F(2) = 8182845904$
$F(3) = 8747135266$
$F(4) = 7427466391$
$F(5) = \underline{\hspace{2cm}}$

Finally, those who found the value of $F(5)$, and went to the site shown, would get a message inviting them for an interview:

> Congratulations. Nice work. Well done. Mazel tov. You've made it to Google Labs and we are glad you are here.

The irony is that many people who hadn't solved the problem themselves were able to Google the answer on sites where it had been posted – though it's doubtful this helped them get much further in the interview process!

We could hardly have a list of the top seven numbers of all without including the very first, the one and only ... one. Because any number multiplied by 1 stays the same, 1 is its own factorial ($1! = 1 \times 1$), its own square, and its own cube; it's also its own inverse, $1/1 = 1$. One is the first odd number, the first positive natural number, and the only number to be neither composite (that is, having factors other than itself and 1) nor prime. It's the first triangular number, or figurate number of any kind, and the first and second member of the Fibonacci series (a series obtained by starting with 1 and then adding together the previous two numbers: 1, 1, 2, 3, 5, 8, 13 ...) As a recurring decimal, 1 can be written as 1.000... or, less obviously, as we've seen, as 0.999...

One plays a crucial role in fundamental areas of maths, such as set theory and the axiomatisation of number systems.

In the standard set of rules, or axioms, generally accepted as the basis for the system of natural numbers, 1 serves as the 'successor' of 0, in other words, the vehicle for generating the next member of the set.

In the realm of philosophy, the One is often taken to be the true or ultimate state of reality. The multiplicity of what we see, according to this viewpoint, is an illusion, and everything in the final analysis is part of an undivided, interconnected whole. Physics, by and large, agrees with this concept, since nothing in nature, due to interactions such as gravity, can be thought of as existing in isolation. What's more, according to cosmologists, all of the matter and energy in the universe is regarded as having originated some 13.8 billion years ago in a single event and at a single moment and point. The Pythagoreans held a broadly similar view: that all of creation stemmed from the monad – the first thing that came into existence – which gave rise to the dyad, which in turn was the source of all the numbers.

The number 1 was conceived long before the number -1. Like zero, negative numbers had to be invented (or discovered) because it isn't obvious, to start with, that they need to exist. After all, you can't have -3 sheep or -8 loaves of bread. But they did prove useful in time, both in maths for its own sake and in practical everyday affairs. Then, after many more centuries, mathematicians began to wonder about the square roots of negative numbers. Everyone knows that the square root of 25 is 5. But what number times itself gives -25? In other words, what's the solution to the equation $x^2 = -25$? The answer couldn't be a real number – a number on the line that stretches away endlessly towards greater positive numbers and, in the opposite direction, towards greater negative numbers. It must be some new beast, a type of number never previously encountered. While some mathematicians in the seventeenth century, and even earlier, began to take seriously the possibility that there could be numbers whose square was negative, others ridiculed the idea and called such numbers 'imaginary'. The name stuck, even though it's misleading, and $\sqrt{-1}$ is still known as the imaginary unit, or simply i.

It's easy to see why π, e, and 1 belong in our top seven because they're so common in both maths and the real world, and also because they're positive numbers that we can measure and deal with in the usual way. But i, at first, doesn't seem to qualify for any kind of accolade. Few of us ever come across it unless we take an advanced maths course in school or go on to specialise in maths or physics, and it's never something we encounter in our daily lives. Yet, for all that, i is something very special. First, it's the basis for a whole system of numbers that massively extends the real numbers. The discovery of this system, known as complex numbers (again a misnomer!), opened up a vast new realm of mathematics, equivalent to astronomers finding that there's a universe of inconceivable size lying beyond the solar system. i is the building block of complex numbers – numbers, like 5 + 2i, that have both a real and an imaginary part. Complex numbers are the basis for complex analysis (the study of functions of complex numbers), which, in turn, has delivered crucial breakthroughs in number theory, algebraic geometry, and many areas of applied maths.

Modern physics would hardly be possible without i. The fundamental equation in quantum mechanics – the Schrödinger equation – has i in it, and the solutions to the equation, known as wavefunctions, are complex numbers. Even in classical physics, whenever there's a need to model periodic motion of some kind, such as that of water waves or light waves, i makes an appearance. Real numbers alone may be fine for describing idealised situations, such as pendulums that swing forever. But as soon as complicating factors, such as the friction that damps a pendulum's motion, are included, the best way to deal with the problem mathematically is to let i into the equations. The same is true when trying to solve problems in fluid dynamics, as when the movement of a fluid starts to become unstable and moves towards the point of turbulence. In Einstein's theory of general relativity, a time interval can be thought of as a distance multiplied by i. In electrical engineering, i is used whenever there's a need to represent the amplitude or phase of an alternating current – except that electrical engineers prefer to use j rather than i to represent the square root of -1 to avoid confusion with the symbol for current.

So far, in our septet of numbers that rule the universe we've included π, e, 1, and i. Zero, too, makes it into the hall of fame – for the many reasons described in Chapter 2, which we don't need to go into again here. Astonishingly, all five of these numerical superstars appear together in a single formula known as Euler's identity:

$$e^{i\pi} + 1 = 0$$

This enigmatic equation links five of the most important numbers in maths with four basic operations (addition, multiplication, exponentiation, and equality) in just about the most simple way imaginable. American physicist Richard Feynman called it 'the most remarkable formula in mathematics'. After proving the identity during a lecture at Harvard, nineteenth-century philosopher and mathematician Benjamin Peirce said, 'We cannot understand it, and we don't know what it means, but we have proved it, and therefore we know it must be the truth.'

In fact, proving Euler's identity isn't very hard – it involves just some straightforward arithmetic and calculus using complex numbers. The exponentiation part of the formula turns out to have an elegant geometric interpretation in terms of the movement of a particle – a mobile point – in the complex plane. The exponential function serves to transport a particle starting at 1 across the complex plane with a velocity equal to its distance from the origin, so if the particle is further away then it will move faster. Applied to real numbers, the particle will simply move further and further away from the origin at a faster and faster pace, reaching a value of e^t at any time t. But for imaginary numbers, the particle's velocity is at 90 degrees to its position, so the path traced is a circle. It takes a time of 2π to make one full revolution of the circle (as the circle has circumference $2\pi r$). So after a time of π, it is halfway around the circle, at -1. This is another way of explaining how $e^{i\pi} = -1$.

Leonhard Euler was the most prolific mathematician of all in terms of his published output. He worked in a variety of fields and is also regarded as one of the greatest mathematicians who ever lived, so it's not surprising that his name is attached to many results, theorems, and objects in maths on which he did pioneering studies. In this chapter alone we've already come across Euler's number (e) and Euler's identity. Next we come to Euler's constant, the value of which, to five decimal places, was first published by Euler in 1735 in his *De Progressionibus harmonicis observationes*. In 1781 he extended his approximation to 16 digits and nine years after that the Italian mathematician Lorenzo Mascheroni gave it to 32 digits, which is why the number is also called

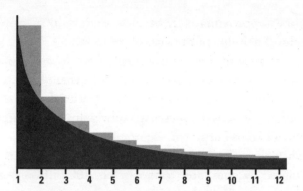

The Euler–Mascheroni constant shown graphically. The light grey area is the difference between $1 + 1/2 + 1/3 + 1/4 \ldots + 1/x$ and $\ln(x)$ (represented by the dark grey area as the integral of $1/x$) as x tends to infinity.

the Euler–Mascheroni constant. Whether the Italian fully deserves this recognition, however, is debatable since he got the last 13 digits wrong! Though less well known than π or e, Euler's, or the Euler–Mascheroni, constant makes it onto our top seven leaderboard for the same reason – the sheer number of places it appears in different areas of maths and its numerous connections with important results and formulae. Euler's constant has come to be denoted by γ (small gamma) because of its intimate link with an important function known as the gamma function – a generalisation of the factorial function – represented by Γ (large gamma). The easiest way to define it is as the value towards which the following expression heads as n gets bigger and bigger:

$$\gamma = 1 + 1/2 + 1/3 + 1/4 \ldots + 1/n - \ln(n).$$

Here ln means natural logarithm. $\ln(n)$ is the power to which e has to be raised to equal n. For example, if $n = 1000$, $\ln(n)$

= 6.908 (approximately), because $e^{6.908} = 1000$ (approximately). The value of the series $1 + 1/2 + 1/3 + 1/4 \ldots + 1/n$, which is called the harmonic series, increases very slowly as n increases, although it does diverge (in other words, grows without limit). The same is true of $\ln(n)$. γ just happens to be the difference between these two slowly diverging functions as n tends to infinity.

The value of γ, which starts off $0.57721566\ldots$, has been calculated by computer to more than 100 billion decimal places. Surprisingly, though, we don't know what kind of number γ actually is. Real numbers may be either rational or irrational, and irrational numbers may be either algebraic or transcendental. We're certain that, for example, 2, 3.14, and 1/3 are rational and we're equally sure that π, e, and $\sqrt{2}$ are irrational. We know, too, that π and e are both transcendental, whereas $\sqrt{2}$ is algebraic. But, strange to say, for all its importance and ubiquity, we don't even know if γ is rational or irrational, let alone whether it's also transcendental. In fact, establishing the status of γ is a major unsolved problem in maths. David Hilbert thought the problem, in his day, was 'unapproachable'. Two giants of number theory, the British mathematicians John Conway and Richard Guy, have said they are 'prepared to bet that it is transcendental'. All we can say for sure, at the moment, is that if γ is rational, in other words can be written as a/b where a and b are both whole numbers, then b must be at least 10^{242080}.

There's a similar constant to γ that applies specifically to prime numbers and is called the Meissel–Mertens constant. Look at this series:

$$N = 1/2 + 1/3 + 1/5 + 1/7 + 1/11 \ldots + 1/n - \ln(\ln(n)).$$

The Meissel–Mertens constant, M, is defined as the limit of N as n tends to infinity. In other words, it's the value towards which this series approaches as n gets bigger and bigger. The sum of reciprocals of primes diverges incredibly slowly as shown by the fact that the difference between it and $\ln(\ln(n))$ is only $M =$ approximately 0.2615. Although it's known that $\ln(\ln(n))$ diverges to infinity, you would never guess it by the rate at which it grows. In fact by the time n has reached a googol, 10^{100}, $\ln(\ln(n))$ is a mere 5.4 or so. When n has climbed to the dizzying heights of a googolplex (10^{googol}), a number so gargantuan that there isn't space enough in the universe to write it out even if the zeroes were written as small as quarks, $\ln(\ln(n))$ is still only around 231.

γ itself turns up in many places in number theory, analysis (of which calculus is a part), and the manipulation of functions. Although obscure to most of us, these appearances by γ are of vital concern to both mathematicians and scientists. For example, γ is central to something called the Gumbel distribution, which can be used to predict future maxima and minima if previous extreme values are known. This makes it of great practical value in predicting the chances of a natural disaster, such as a volcanic eruption or an earthquake, occurring in a given period of time. Through its role in the gamma function Γ, mentioned earlier, γ is involved in the modelling of cryptographic systems and therefore in the maths of ensuring secure transactions. γ also shows up in the solutions of Bessel functions, which can be used to model wavelike systems, including the design of waveguide antennae, the vibration of membranes, and the conduction of heat through substances – all problems relevant to the design of mobile phones.

Last, and certainly not least, in our Big Seven list is a number that we have no way of imagining. Philosophers have long pondered on the infinite but mathematicians tended to shy away from it. They were happy to acknowledge, for instance, that there was no end to the numbers and that lines could extend indefinitely in any direction. But they were reluctant to deal with infinity as a mathematical thing – until Georg Cantor came along and, despite fierce opposition, established set theory and the existence of different orders of infinity.

Cantor called the smallest kind of infinity, which is the size of the set of all natural numbers, aleph-null ($\aleph_0$), aleph being the first letter of the Hebrew alphabet. It's the first of what are known as transfinite numbers. You may sometimes hear it said that infinity isn't a number but the truth is it's a different kind of number. Transfinite numbers obey strict rules and behave in ways that can be known and analysed. It's just that their behaviour is completely different from anything else with which we're familiar.

Adding any number to $\aleph_0$ doesn't change it. $\aleph_0 + 1 = \aleph_0$. $\aleph_0 + 1000 = \aleph_0$. You can even add $\aleph_0$ to itself, or multiply $\aleph_0$ by any finite number, and it will still be $\aleph_0$. It seems impregnable, aloof. But there is a way to jump to a different kind of infinity and that is to use $\aleph_0$ as a power. As soon as we write $2^{\aleph_0}$, or in fact raise any finite number or even $\aleph_0$ itself to the power $\aleph_0$, we move up the hierarchy of infinities to aleph-one, $\aleph_1$ (assuming that something called the 'Continuum Hypothesis' is true – about which we'll have more to say in Chapter 13). Mighty though $\aleph_0$ is, it's just the first of infinitely many infinities, each one infinitely larger than the one before. The mind boggles – but then it's bound to, because it's finite.

$\aleph_0$ makes it into our list of the seven greatest numbers, not just because of its size but also because it's representative of a genuinely important kind of mathematical object. Anyone who comes across the limits of series in school maths or does some introductory calculus will come across infinity. In fact, the concept of infinity underpins the whole field of real analysis, which forms the foundation of calculus. It's also at the core of measure theory from which stems our deepest insights into problems related to probability. In physics, Hilbert spaces, used in the formulation of quantum mechanics, are infinite not just in size but also in dimension. Finally, lying far beyond $\aleph_0$, but derived from it, are exotic transfinite numbers that find application in the most foundational issues of mathematics and in generating, through a function called the 'fast growing hierarchy', the largest finite numbers ever contemplated by the human mind.

CHAPTER 4

Through the Looking Glass

> Symmetry, as wide or as narrow as you may define its meaning, is one idea by which man through the ages has tried to comprehend and create order, beauty and perfection.
>
> – Hermann Weyl

ONE OF US (David) used to have a school maths teacher – Mr Kaye – who had a favourite question for his older students. 'How,' he'd ask, 'did asymmetry get into the universe? That's what I want to know.' It's a puzzle almost as basic as why there's something rather than nothing: why is there asymmetry rather than symmetry? Or, to put it another way, how did the universe come to distinguish between one side (of anything) and another?

Symmetry and asymmetry coexist in almost everything. The human body, seen from the front or back, is externally more or less bilaterally symmetric. Faces are perceived as being more attractive if they're symmetric, but most are surprisingly asymmetric. Internally, the body is a mixture of symmetry and asymmetry. Such, too, is the case in maths and nature.

In mathematics, and in life, the first examples of symmetry we encounter are those in the objects around us and in geometry. Some things, we notice, are the same on one side as the other. This is what's known as line symmetry: when a shape is identical to its mirror image. Printed upper-case letters are examples of shapes with different lines of symmetry. Some letters, like a capital M (in a sans-serif font) or C, have one line of symmetry. Others, like G, have none. H has two lines of symmetry, one vertical and one horizontal. Two interesting examples are X and Y. The former, as shown printed here, has only two lines of symmetry but it can be written in such a way as to have four (when the angles between the arms of the X are all 90°), in which case the symmetry lines are the horizontal and vertical bisectors, and both diagonals. The Y shown here has only one line of symmetry but can be rewritten to have three (when the angles between the arms are all 120° and all arms are of equal length).

Mirrors are fascinating when it comes to thinking about symmetry. Why does a mirror reverse right and left, but not up and down? This question crops up perennially in the letters and queries columns of magazines and newspapers, and was the inspiration for Lewis Carroll's *Through the Looking-Glass*. One day in late 1868, a girl called Alice Raikes was playing in the garden of her home in Onslow Square, London, which was adjacent to where Charles Dodgson (Lewis Carroll) used to stay with his uncle. One day, he called over to her: 'So you are another Alice. I'm very fond of Alices [his famous books are named after Alice Liddell, the daughter of the dean of Christ Church college, Oxford]. Would you like to come and see something which is rather puzzling?' She followed him into his uncle's house, to a room with a tall mirror standing across one corner. He then asked her to hold an orange.

'Now, first tell me which hand you have got that in.'

'The right.'

'Now, go and stand before that glass, and tell me which hand the little girl you see there has got it in.'

'The left hand.'

'Exactly, and how do you explain that?'

'If I was on the other side of the glass, wouldn't the orange still be in my right hand?'

'Well done, little Alice. The best answer I've heard yet.'

Recalling this conversation years later, Alice Raikes (Mrs Wilson Fox) said: 'I heard no more then, but in after years was told that he said that had given him his first idea for *Through the Looking-Glass*, a copy of which, together with each of his other books, he regularly sent me.'

Alice enters the looking glass. Artwork by John Tenniel.

Returning to our question, if a mirror swaps right and left, why not also up and down? A frequently given answer is that a mirror doesn't reverse right and left. It reverses *front and back*, which is certainly true: your reflection in the looking glass faces in the opposite direction to the real you. But this short explanation doesn't completely dispel the mystery. The fact is that, if you imagine that the mirror weren't there and that instead you were looking at a flesh-and-blood twin of yourself, that twin would be differently handed. If you have a watch on your left wrist, the person facing you has his/her watch on the right wrist. The mirror *has* done a left-right swap, surely! At any rate, something has happened to left and right that hasn't happened to up and down. To be more convinced of this, hold this book up to the mirror and try to read it. If no left-right swap has happened, why is the reflected writing so hard to read? Firstly, remember that you're only looking at an image. The mirror hasn't (Carrollian fantasies aside) created some*thing* of opposite handedness. Secondly, appreciate, as Alice Raikes might have done, how the writing appears in the mirror's frame of reference. This is easy to do by looking at the writing from the other side of the page (i.e. back to front, thus undoing the back to front reversal caused by the reflection). From the mirror's point of view the writing looks perfectly normal.

Nature does a fair amount of mirror reversal. In the case of identical twins it can happen when the split takes place more than a week after conception (but not so late as to result in conjoining). The mirror imaging may take the form of opposite hair whorls, right handedness in one twin contrasted with left handedness in the other, teeth that erupt on opposite sides, leg-crossing in opposite ways, and, in extreme cases, organs that are reversed left to right. DNA tests would reveal no differences between mirror-image twins, only that they were identical. The double spiral shape of DNA molecules always coils in the same direction. But there are many organic (carbon-based) molecules that *do* have right and left forms. In chemistry, this is a property called chirality. The mirror images of a chiral molecule are known as enantiomers or optical isomers, and individual enantiomers are said to be right-handed or left-handed. Chemists can distinguish between right- and left-handed versions of a substance by passing plane-polarised light (light that vibrates in just one plane at right angles to its direction of travel) through it. Right-handed, or dextrorotatory, molecules rotate the plane of polarisation to the right, and left-handed, or laevorotatory, molecules rotate it to the left.

Many biologically important molecules are chiral, including sugars and the naturally occurring amino acids, which are the building blocks of proteins. Most sugars found in living things on Earth are dextrorotatory (D), whereas most amino acids are laevorotatory (L). Interestingly our taste and smell (olfactory) receptors are chiral so that they respond differently to the L- and D-forms of molecules. L-forms of amino acids, for instance, tend to be tasteless, whereas D-forms taste sweet. A chemical called carvone is found in both spearmint leaves and caraway seeds, but the smell and taste of each differs greatly because our taste buds and nasal receptors respond quite differently to the L-enantiomer of carvone found in spearmint and the right-handed form in caraway.

Line or mirror symmetry, also known as reflection symmetry, is just one of the forms that symmetry can take in geometry. Another is rotational symmetry, in which a shape remains the same after a partial turn about a point (in 2D) or an axis (in 3D). The rotation that brings the shape back into itself may be 180°, 120°, 90°, or, any value 360°/n, where n is a whole number. Rotational symmetry can exist on its own, without any line symmetry. For example, the letter N has rotational symmetry of order 2, which means it remains the same after a 180° turn about its centre. But a shape with two or more lines of symmetry must necessarily also have some rotational symmetry as well. In particular, any shape with exactly n lines of symmetry must necessarily have rotational symmetry of order n (it remains the same when rotated by 360°/n).

The letter O is an interesting case. Represented in this fashion, as an oval, it has only two lines of symmetry and thus also rotational symmetry of order 2. Yet if the O is written as a perfect circle, something intriguing happens: it has infinitely many lines of symmetry (indeed, any line through the centre is such a line) and can be rotated by any angle whatsoever about the centre and remain the same. Any such shape in the plane that has this symmetry must consist entirely of concentric circles.

In the geometry we do at school, line symmetry and rotational symmetry crop up a lot. But we don't hear much of other types. One of these is translational symmetry, which occurs when a shape remains the same when moved in the plane. A honeycomb pattern displays translational symmetry – in three different directions – because it's made up of regular hexagons, all of the same size and orientation, which fit together snugly like a jigsaw puzzle. Of course, real honeycombs are finite in size, whereas translational symmetry is a characteristic of only infinite patterns, so we have to imagine the honeycomb extending forever in all directions. A straight line also possesses infinite translational symmetry when moved along itself. The difference between this case and that of an infinite periodic tiling, such as a honeycomb pattern or a square tiling, however, is that the latter is discrete; in other words, the tiling must be translated by a multiple of a fixed amount, rather than being allowed to be translated by any distance whatsoever, as is the case with a straight line.

A fourth type of symmetry is known as glide reflection symmetry. A glide reflection consists of reflecting a shape in a line and then moving it along in the same direction as the line. Like translational symmetry, this can't occur in finite shapes – indeed, whenever a shape exhibits glide reflection symmetry it must be part of an infinite pattern and also have translational symmetry.

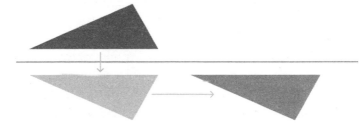

A glide reflection: a combination of reflection across a
line and a translation parallel to the line of reflection.

All geometric symmetries in the plane (provided the symmetry is *rigid*, which means that it mustn't bend or stretch the plane) fall into one of these four categories – reflection, rotation, translational, and glide reflection. Yet in three or higher dimensions there are many other symmetries, such as central symmetry (where the shape is reflected in a point rather than in a plane) and screw symmetry (a rotation about an axis followed by a translation along the axis, like a screw).

With the existence of so many different types of symmetry, especially in three dimensions or more, the question naturally arises of whether there's some neat and effective way to classify them all for a given object. In fact there is but it takes us away from the familiar reflecting and rotating of shapes in elementary maths and into a much more abstract realm called group theory. One of the slightly confusing aspects of maths is that it gives unexpected and very specific meanings to a lot of well-known words, such as 'irrational', 'imaginary', 'set', 'field', and 'ring'. To a mathematician a group is a set, or collection of items, that share the same multiplication table. What this means is that for any two elements a and b in the group, there's another element $a \cdot b$, which is essentially 'a multiplied by b'. The multiplication can't be completely arbitrary – it must satisfy some properties for the set to count as a true group. First, it must be associative, that's to say the product of three or more elements doesn't depend on where the brackets are placed, so $(a \cdot b) \cdot c = a \cdot (b \cdot c)$. Second, there must be an identity element, e, such that $a \cdot e = a$ and $e \cdot a = a$. So, e behaves like 1 does when we're doing ordinary multiplication or 0 when we're doing ordinary addition; in either case, the identity element doesn't change the result. Finally, for every a there must be an inverse, usually written as a^{-1}, so that $a \cdot a^{-1} = a^{-1} \cdot a = e$.

Notable by their absence are other properties that we're accustomed to in conventional arithmetic, such as the order in which things are multiplied. We're familiar with the idea, for example, that $2 \times 3 = 3 \times 2$. But in the case of groups, it's not always the case that $a \cdot b = b \cdot a$. (A group for which a • b *is* always equal to b • a is a special kind, which we'll come back to later, known as an abelian group.) From the basic group properties we can deduce certain important things. For example, for any two elements a and b, there must be another, unique element c such that $a \cdot c = b$.

Now that we've got all the essential machinery in place needed to deal with groups, we can take a look specifically at symmetry groups. Given a shape, we can construct the set of all its symmetries – all the transformations that don't change the appearance of the shape. The identity e is simply the 'transformation' that changes nothing. The '·' is taken to mean apply one action followed by another; $a \cdot b$, for instance, is shorthand for 'perform b then a'. So, if a is 'reflect in the y-axis' and b is 'rotate by 180°', then $a \cdot b$ means 'rotate by 180° and then reflect in the y-axis', which turns out to be the same as the single action 'reflect in the x-axis'. Finally, the inverse a^{-1} simply means 'perform a in reverse', so if a is 'rotate clockwise by 60°', a^{-1} is 'rotate anticlockwise by 60°'.

The simplest symmetry group of all is that of a completely asymmetric object, such as the letter Я on a sheet of paper. There's just one element of the group, namely, the identity element e, and the multiplication table for the group consists solely of $e \cdot e = e$. The only symmetry in this case involves doing absolutely nothing, and, for obvious reasons, the group is known as the trivial group.

Of shapes with nontrivial symmetries, some are described by so-called cyclic groups. The cyclic group of order n can be understood in many ways, but it's equivalent to the group of integers modulo n (which simply means the remainders when dividing an integer by n) where · corresponds to addition and e corresponds to 0. Modular arithmetic is often known as clock arithmetic, because a clock gives a good example of it. An analogue clock works modulo 12: if it's seven o'clock and you add eight hours, it becomes three o'clock (not fifteen o'clock). The analogue clock is equivalent to the group of integers modulo 12. Digital clocks generally use 24-hour time and so represent the integers modulo 24. Any shape in the plane that has rotational symmetry of order n but no lines of symmetry or translational symmetry will have as its symmetry group the cyclic group of order n, denoted Z_n.

Another type of symmetry group is the dihedral group D_n. It's the symmetry group of a flat shape that has n lines of symmetry and, therefore, rotational symmetry of order n. For instance, it's the symmetry group of a regular polygon with n sides. A dihedral group D_n has $2n$ members: n rotations plus n reflections. Unlike cyclic groups, dihedral groups aren't abelian: $a \cdot b$ isn't necessarily the same as $b \cdot a$. To see this, suppose a and b are two reflections of an equilateral triangle. Two reflections, one after the other, will result in a rotation, but if the order of reflections is reversed, the rotation will be in the opposite direction.

Cyclic groups and dihedral groups are the only finite two-dimensional symmetry groups. Any shape that has translational symmetry will have an infinite symmetry group. In 3D space, however, due to the richer variety of symmetries, more complex groups are possible. The symmetry group of a tetrahedron, for instance, is S_4 – the group of all permutations of four objects. One such permutation might be 'rearrange $\{1, 2, 3, 4\}$ to $\{2, 4, 1, 3\}$'. Another might be 'rearrange $\{1, 2, 3, 4\}$ to $\{1, 3, 2, 4\}$'. The identity is, again, just 'do nothing'. To see that this group is the same as the symmetry group of a tetrahedron, notice that if we label each vertex of a tetrahedron with a number, we can, simply by rotating and reflecting, rearrange these numbers into any order we choose.

We're used to thinking of symmetry in terms of shapes, objects we can see or imagine, and geometry as a whole. Yet it's a concept important in other fields of mathematics, such as algebra and, in particular, polynomial equations. An equation of this type has terms that involve powers of x, for example, $x^5 + 3x^4 - 2x + 8 = 0$. In algebra, it's often required that the coefficients (here 1, 3, 0, 0, -2, and 8) of the polynomial are all integers, as otherwise any real number could be a solution of the equation. The numbers that occur as solutions to a polynomial are known as algebraic numbers. All rational numbers are algebraic, as is, for instance, the square root of 2 but not π (which is, instead, transcendental). Algebraic numbers can themselves have symmetries: sometimes there are cases where the appearance of one number as a solution of a polynomial forces the appearance of a symmetric co-solution, as in the case of the pair $1+\sqrt{2}$ and $1-\sqrt{2}$. Every polynomial that has $1+\sqrt{2}$ as a root (for instance $x^2 - 2x - 1 = 0$) also has $1-\sqrt{2}$ as a root.

With linear equations (where the highest power of x is 1) solving the equation is a trivial matter. For example, $4x + 3 = 0$ has the single solution $x = -3/4$. With quadratic equations, the solutions can always be found using the quadratic formula: if $ax^2 + bx + c = 0$, then:

$$x = \frac{-b \pm \sqrt{b^2 - 4ac}}{2a}$$

The plus/minus sign ($\pm$) here means that we can use either a + or a − . Both will work and give two distinct values of x that satisfy the original equation (unless $b^2 - 4ac = 0$, in which case the values of x are the same).

If $b^2 - 4ac$ has a negative value, then x will be a complex number, consisting of a real number plus a multiple of i, the square root of -1. Every complex number $a + bi$ has a corresponding complex conjugate, $a - bi$, and, it turns out, that whenever one of them is a solution to a polynomial equation, so also is the other. You might think that the discovery of complex numbers came about in this way, through attempting to solve quadratics for which $b^2 - 4ac$ is negative. But in fact mathematicians, for quite a while, were happy to assume that such equations simply had no solutions.

In time, the question naturally arose: could polynomial equations other than the linear and quadratic types be solved? During the Renaissance, competitions between pairs of mathematicians were common. Each would set problems for the other to solve and the contestants would often bet on being able to win. One such competition was between the sixteenth-century Italian mathematicians Niccolò Tartaglia and Antonio Fiore. Niccolò's family name was Fontana but he was nicknamed Tartaglia (meaning 'stutterer') because of a speech impediment that resulted from a sabre cut to his jaw and palate that he received when the French sacked his hometown of Brescia when he was just a boy. Tartaglia knew that Fiore had learned how to solve one of three classes of cubic equations – polynomials in which the highest power of x is three. Tartaglia, however, could solve all three, so to ensure that he won the competition he set questions to which he knew the answers but that would baffle Fiore. Afterwards, in 1539, another Italian mathematician, Girolamo Cardano, persuaded Tartaglia to reveal his method for solving cubic equations, on the condition that he kept it secret. Tartaglia was then incensed to discover, a few years later, that Cardano had published a detailed description of the method in his *Ars Magna* (The Great Art). There for all to see was Tartaglia's general cubic formula, which could solve cubic equations just

Niccolò Tartaglia. About the events that led to his nick-name 'the stutterer', he wrote: 'In the cathedral, in front of my mother, I was given five murderous wounds... One of the wounds cut my mouth and teeth, breaking my jaw and palate in half. This stopped me from talking except in my throat the way magpies do.'

as the well-known quadratic formula could solve quadratics. It later transpired that Cardano had found out about the work of another mathematician, Scipione del Ferro, Fiore's teacher, who'd also managed to solve the cubic equation, and thus he could claim that his method in *Ars Magna* originated with someone other than Tartaglia.

Cardano realised that, in some cases, del Ferro's method required finding the square root of a negative number. His immediate instinct was to declare that such equations had no valid solutions. But ignoring the problem didn't make it go away. There were cubic equations, such as $x^3 - 15x - 4 = 0$ that, in solving by del Ferro's method, invoked the square roots of negative numbers, yet nevertheless ended up with real solutions. Given that even negative numbers were, at this time, still treated with suspicion, any talk about the square roots of negative numbers must have seemed like utter madness. In *Ars Magna*, Cardano showed how to manipulate such weird square roots, but it was clear that he didn't consider them to be genuine numbers. Instead he treated them as mere conveniences – handy stepping-stones in getting to the correct answer. It wasn't until 1572, when Rafael Bombelli published his *Algebra*, that imaginary numbers were granted full mathematical citizenship and treated as meaningful entities in their own right.

Earlier, in 1540, even before the publication of *Ars Magna*, one of Cardano's students, Lodovico Ferrari, managed to find a solution to quartic equations (polynomial equations whose highest power is x^4). The inclusion of Ferrari's quartic formula in *Ars Magna* encouraged the quest for a solution to the quintic equation (with highest power x^5). However, this proved to be a much tougher nut to crack, and in time it became clear why.

In 1799, the Italian mathematician and philosopher Paolo Ruffini published a proof that there's no general method for solving quintics. It turned out that his argument, while mostly correct, contained one major gap. Fortunately, a quarter of a century later, the Norwegian mathematician Niels Henrik Abel was able to fill this gap and provide a complete proof that there's no one-size-fits-all formula for finding the solutions of quintic equations. Abel's proof involved groups of a type that we mentioned earlier and that are named after him – abelian groups. While his proof firmly slammed the door on any chance of ever finding a general quintic formula, it still left open another possibility: that quintic equations might all be solved on a case-by-case basis.

In Chapter 8, we'll come across a colourful character and mathematical genius by the name of Évariste Galois. Tragically, he died at the age of only twenty having rashly agreed to take part in a pistol duel. But the night before, sensing that his end was near, he desperately tried to write down his most important mathematical discoveries. Out of these last-minute scribblings, preserved in the form of letters to friends, arose the field of Galois theory.

Galois was interested in the symmetry of polynomials. To every polynomial he assigned a group, now known as a Galois group. The Galois group of a polynomial describes how you can rearrange its solutions in such a way as not to change which other polynomial equations remain the same. For instance, a polynomial such as $x^2 - 3x + 2 = 0$ has two solutions, $x = 1$ and $x = 2$. No rearrangement is possible (as certain polynomials, such as $x - 1 = 0$, have only $x = 1$ as a solution, not $x = 2$) so the Galois group is the trivial group, with only one element. A different quadratic, $x^2 - 2x - 1 = 0$, has two solutions, $x = 1 + \sqrt{2}$ and $x = 1 - \sqrt{2}$. Yet in this case, the different values can indeed be rearranged. This even preserves polynomials in more than one variable; for instance, if we let the former solution be a and the latter be b, $a + b = 2$ and this stays the same upon rearrangement. The Galois group is the cyclic group of order 2, which is also the symmetry group of a letter like **M**.

In the case of quadratics, these are the only two Galois groups and both are very simple. But the young Frenchman realised that polynomials of higher degree can have more interesting Galois groups. In a key discovery, he showed that the Galois group of a particular quintic was the group of rotational symmetries of a dodecahedron, and that any polynomial described by such a group couldn't possibly be solved using only standard operations and roots.

Group theory has come a long way in the nearly two centuries since Galois died. One of the great achievements in it has been the classification theorem of finite simple groups. A simple group is analogous to a prime number in that it doesn't have any normal subgroups other than itself and the trivial group. The theorem states that every finite simple group belongs to one of 18 categories, or is one of 26 'sporadic' groups, which don't fit any pattern. The simplest category is that of the cyclic groups of prime order. These are just the groups modulo p under addition, where p is a prime number (the same as the group of a clock with p hours).

The next simplest category is that of alternating groups of degree greater than or equal to 5. Suppose you had n numbers and could, in one step, swap any pair of them. If you could do this unrestrained, you could make any permutation of those n numbers, and the group formed would be the symmetric group of degree n. But if you were restricted so that you always had to make an *even* number of swaps, you'd end up with the alternating group. For instance, the permutation that rearranges $(1, 2, 3, 4, 5)$ to $(2, 1, 4, 3, 5)$ is in the alternating group of degree 5, but the permutation that rearranges it to $(1, 3, 2, 4, 5)$ is not, because an odd number of swaps (here, one) has been used. The alternating group of degree 3 turns out already to have been classified (it's the same as the cyclic group of order 3) and the alternating group of degree 4 is a special case – it isn't simple as it contains the dihedral group D_2 as a normal subgroup. But these exceptions aside, all alternating groups are simple. (For example, the alternating group of degree 5 is the group of rotational symmetries of the dodecahedron.)

The remaining 16 categories of groups are a lot more complicated than either of the two just mentioned, and are collectively referred to as the groups of Lie type, after the Norwegian mathematician Sophus Lie. Falling outside any of the 18 categories are 26 other groups – the sporadic groups – which defy all attempts at classification. Five of them were discovered by Émile Mathieu, the first in 1861, and are named after him. By far the largest of the sporadic groups was found by the American mathematician Robert Griess in 1976. Known as the Monster group, it has just over 808 thousand trillion trillion trillion trillion elements, which Griess managed to express using 196,883 by 196,883 matrices. Nineteen of the other 26 sporadic groups, it turns out, are related to the Monster, and together with the Monster form what Griess called the happy family. The six remaining outcasts go by the less happy name of 'the pariahs'.

The classification theorem of groups involved a decades-long collaborative effort that culminated in a truly gigantic proof. Although still possible to verify by hand, it was between 5,000 and 10,000 pages long and spanned approximately 400 journal articles. Being a project that involved many mathematicians, no one person kept track of everything so that the true size of the proof remains vague.

The groups we've talked about mainly so far – the finite groups – apply to the symmetry of mathematical or physical objects when just a finite number of transformations are possible that preserve the object's structure. But there's a whole other class of groups that apply to *continuous* transformations. These were first studied by Sophus Lie at the end of the nineteenth century and so bear his name. Lie groups, confusingly, aren't the same as Lie type, which, as we've seen, are finite groups! Instead, Lie groups deal with objects that preserve their appearance under continuous transformation. A simple example is a sphere, which looks the same no matter how much it's rotated.

Lie's main interest was in solving equations. When he started his research, the methods available for solving equations formed a sort of bag of tricks. A typical solution technique was to make a clever change of variables, which would cause one of the variables to drop out of the equations. Lie's crucial insight was that when this happened it was due to an underlying symmetry of the equations – a symmetry that he was able to capture in a new type of group.

The theory of groups, finite and continuous, is now of immense importance in both maths and science. It was used, early on, to determine the possible structure of crystals, and has come to have deep implications for the theory of molecular vibration. Group theory has permeated the work of physicists studying the elementary particles and forces in nature, and one of the simplest groups, the multiplicative group modulo n, is used every time you send secure information over the Internet.

Crystals can come in many different structures, and their symmetry groups can help us to understand more about how they form. Halite crystals, for example, consist of sodium and chlorine ions arranged in a cubic lattice, so that they have all the symmetries of a cube plus an infinite set of translational and other related symmetries. What's interesting is that crystals of the same substance will have certain properties that don't immediately seem obvious from their microscopic structure. For instance, the angles between two faces of a particular type will always be constant, regardless of the size or overall shape of the crystal. Halite crystals aren't always perfect individual cubes – they often appear to consist of many overlapping cubes stuck together – but the angle between two faces will always be 90°. This description of the angles is known as the crystal habit, and from a crystal habit a symmetry group of the microscopic structure can be determined. Different crystals may have different habits. The habit of diamond, for example, is the face-centred cubic lattice. This turns out to be the most efficient way to pack atoms, a fact that underpins diamond's status as the hardest natural material.

The conservation laws of physics, such as conservation of energy and conservation of electric charge, all arise from symmetries in the underlying equations. It used to be thought that there were three types of fundamental symmetry of the universe: those of charge, parity, and time. Charge symmetry simply means that the laws of physics remain the same if all matter is exchanged with antimatter and vice versa.

A cluster of quartz crystals.

Parity symmetry essentially means that the laws of physics don't distinguish between left and right, so if the entire universe were reflected in a mirror all laws would stay the same. Finally, time symmetry means that the laws of physics are unchanged if we reverse the direction of time.

The last of these symmetries seems at first glance very counterintuitive. It suggests, for instance, that if a vase were to fall off a shelf and break, the laws of physics would allow it to spontaneously reassemble and hop back on the shelf. In fact, that's possible – though fantastically unlikely. The reason it doesn't happen lies with the second law of thermodynamics, which is more statistical in nature rather than physical. It deals with a quantity known as entropy, which is a measure of disorder. Entropy relates to the number of microstates of a system – the number of ways in which it can be rearranged while still preserving its physical properties. For instance, a pack of cards in pristine order (ace to king in each suit) has extremely low entropy as there's little that can be changed in the system while preserving the fact that the cards are in optimal order. At most, you could swap the suits, but this gives only 24 microstates. By contrast, a randomly shuffled pack of cards has high entropy because it can be rearranged in $52 \times 51 \times 50 \ldots \times 2 \times 1$ (greater than 8 followed by 67 zeroes) ways to give any other randomly shuffled sequence. The second law of thermodynamics states that the total entropy will always increase (at least in the case of an isolated system, where neither matter nor energy can enter or leave). It's theoretically possible that entropy can decrease in a particular instance, but with a vanishingly small probability. Shuffling an already thoroughly shuffled pack of cards *could* result in a perfectly ordered pack of cards but it's vastly more likely to result in just another of the trillions upon trillions of disordered microstates. Likewise, a broken vase has a much greater entropy than an undamaged vase and so, while the laws of physics don't actually disallow the vase reforming and jumping back on the shelf, it's overwhelmingly more likely to remain in one of the countless shattered microstates.

So, the second law of thermodynamics explains the apparent asymmetry in time, while in fact leaving the three key symmetries intact. Of the four fundamental forces in nature, three of them – gravity, electromagnetism, and the strong nuclear force – were shown to obey all three basic symmetries (charge, parity, and time). Physicists assumed that eventually the remaining force, the weak nuclear force, would fall into line. But then, in 1956, the Chinese-American physicist Chien-Shiung Wu carried out an experiment to measure the decay of a radioactive isotope of cobalt, cobalt-60, in a magnetic field at very low temperature. As cobalt-60 decays, it emits electrons. Wu observed that these electrons were much more likely to be emitted in a particular direction – opposite to that of the nuclear spin. If parity symmetry held, it should be then in a mirror reflection of our universe, the direction in which electrons were given off should be the same. Yet, Wu had shown that, in fact, it would be reversed.

The fall of parity shocked the physics community. 'That's total nonsense!' said leading theorist Wolfgang Pauli. Others rushed to reproduce the experiment, many certain that Wu must have made a mistake, but all ended up verifying her results. Parity symmetry had indeed been broken. Some physicists fell back on the possibility that CP together – charge-parity symmetry – might be conserved, by suggesting that antimatter is identical to the mirror image of matter (so that anticobalt would behave as the exact mirror image of cobalt in Wu's experiment). Yet, in 1964, CP, too, was shown to be broken. Time symmetry was broken as well, and at the most fundamental level of the elementary particles and fundamental forces, rather than just at the statistical level of the second law of thermodynamics. That left only CPT symmetry as a candidate for the true symmetry of the universe. CPT symmetry states that if you reverse all three of charge, parity, and time, the laws of physics of the universe will be identical. This would mean that, for instance, violations of CP symmetry are identical to violations of time symmetry. According to our current mathematical model of the universe, CPT symmetry is truly unbreakable. Indeed, it's been found to hold in all the experiments we've devised so far. But whether it's a true symmetry of the universe or whether we'll come up with a new theory in which even CPT symmetry is broken is, as yet, unknown.

We've travelled far in our exploration of symmetry but not really solved Mr Kaye's riddle. On the contrary, the problem seems to have multiplied. How did the three symmetries of charge, parity, and time all individually come to be broken, leaving only their combination intact? Why isn't the universe just a homogeneous gas cloud, identical in all directions? Or, more to the point, why didn't equal amounts of matter and antimatter survive at the start of the universe, causing them to have long ago annihilated each other and leaving behind only radiation?

Confirmation that there were irregularities in the very early universe came from a space mission called the Wilkinson Microwave Anisotropy Probe (WMAP), which was launched in 2001 and operated for nine years. WMAP detected tiny asymmetries in the cosmic microwave background, the faint afterglow of the Big Bang itself, corresponding to some regions that were very slightly warmer than others. Once such asymmetries took root, they could grow and eventually lead to the lumpiness of the universe we see today, in which galaxies and clusters of galaxies are separated by spaces largely devoid of matter. But the great mystery remains: how did asymmetry come about in the first place? Without it, we wouldn't be here to ask the question. Yet it's a question of the most profound nature. Why is the universe not perfectly symmetric, and when and how did its asymmetry begin?

An all-sky picture of the young universe created from WMAP data. The image reveals 13.77 billion-year-old temperature fluctuations that correspond to the seeds that grew to become the galaxies. Lighter regions have a slightly higher temperature than darker ones.

CHAPTER 5

Maths for Art's Sake

I am interested in mathematics only as a creative art.

– G. H. Hardy

THERE'S A PASSION and a vital force at the heart of maths that's often best revealed through art and music. Mathematicians and artists are drawn to the same patterns – patterns that are embedded in the physical world. Not surprisingly, then, key developments in maths have come about because of the activities of artists and architects, and mathematics permeates the work of some great pioneers of the visual arts.

One of the first artists to go out of his way to incorporate maths into his creations was the Greek sculptor Polykleitos in the fifth century BCE. Polykleitos made statues in bronze and other materials of heroic figures, some of which survive today in the form of Roman marble copies. Like Pythagoras and his followers, whose teachings may have influenced him, Polykleitos believed that mathematics lay at the heart of everything and was essential to achieving artistic perfection. A statue of an athlete or a god, he thought, should be made up of parts that were all in balance and connected by simple mathematical proportions. To him, the square root of 2 – approximately 1.414 – held the key to this system. His starting point was the length of the last bone, or phalange, of the little finger. Multiplying this by √2 gave him the length of the middle phalange, and by √2 again the length of the third phalange. Taking the length of the whole little finger and multiplying this by √2 gave him the length of the palm, from the base of the little finger to the head of the ulna. And applying this same factor over and over again he arrived at the size of the chest, the torso, and so on, until he had every basic dimension for what he considered to be the ideally proportioned male anatomy. His *Canon*, in which he wrote

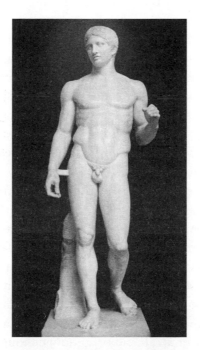

The Doryphoros, marble statue by Polykleitos (*c.* 440 BCE).

down this system of geometrical progressions, served as a guide for the work of many sculptors in ancient Greece and Rome, and all the way through to the Renaissance.

Painters, working on a flat surface, faced the problem of how to portray three-dimensional scenes. Both the Greeks and the Romans wrestled with the task of creating images that had a feeling of depth – and succeeded to some extent. A fresco from the wall of a villa in Pompeii, preserved when volcanic ash engulfed the city in CE 79 and now on display in the Metropolitan Museum of Art in New York, depicts numerous buildings using a system of perspective that does a pretty good job of suggesting depth and distance. Only a closer look at the lines of colonnades and other elements of the picture reveals something not quite right as they stretch away into the distance.

For the most part, artists in the Middle Ages didn't even try to capture accurate 3D views, partly because the older knowledge of how to do it had been lost and partly because the Church – the commissioner and overseer of most works in those days – had no desire to depict things as they really were. It was common, for instance, for mediaeval artists to show people and objects bigger if they were important thematically or spiritually, rather than because of their relative position in the picture.

The breakthrough to a mathematically rigorous form of perspective came in Europe in the early 1400s, at the dawn of the Renaissance. More than a century earlier, the Florentine painter and architect Giotto di Bondone had presaged this with his attempts, using algebra, to show how distant lines in a picture should be placed. But it was his compatriot, the designer and architect Filippo Brunelleschi, who put the subject of what we now call projective geometry on a firm footing.

Brunelleschi was a practical man, trained as a goldsmith, whom some consider to have been the first modern structural engineer. His greatest triumph was to construct a new dome, with an inner diameter of about 150 feet, for the magnificent cathedral in Florence. The cathedral's overseers wanted a dome built of masonry that, despite weighing tens of thousands of tons, would be self-supporting, without the need for flying buttresses and pointed arches that were the only known means of holding up such a massive structure at the time. What's more it had to sit on existing walls that were 180 feet high and octagonal in plan. Brunelleschi won the competition to come up with a design that would meet these extraordinary demands and then set about forging new approaches to building and building site safety, which included providing workers with watered-down wine at lunch

The dome of Florence Cathedral.

to ensure they stayed sober on the job, a safety net to catch anyone who fell, and a chiming clock to mark the change of shifts. To lift the building materials high above the ground, Brunelleschi devised the world's first reverse gear, enabling an ox to raise or lower a load at the flick of a switch.

In 1434, with his fabulous dome nearing completion, Brunelleschi put on a public display of art in which he unveiled another innovation. Earlier, he'd used a mirror to reflect the twelfth-century baptistery of the cathedral, another octagonal structure, and then painted over the reflection to make an exact copy of it. To prove that he'd captured an accurate 2D rendering, he drilled a small hole in the back of the picture then invited visitors to look through the peephole at the real baptistery but with a second mirror reflecting the painted version of the scene. By moving the second, backward-facing mirror in and out of the way, it could be seen that the view of the actual baptistery and its painted image were the same, and both continuous with their surroundings.

The importance of having captured a true perspective view in this way – perhaps for the first time in history – is that Brunelleschi could now scrutinise it with an analytical eye and unpick its mathematical structure. A couple of things stood out as he looked at the lines of the baptistery and neighbouring buildings. First, the central vanishing point, exactly opposite the observer's viewpoint, lay on the horizon line. Second, the horizon passed not only through this point but also through the oblique vanishing points – the lines defining the perspective of the baptistery itself.

Other Renaissance artists, in Italy and beyond, began to incorporate the principles uncovered by Brunelleschi into their own work. Mathematicians used the new knowledge, together with insights of their own, to lay down the foundations of projective geometry. Among the first of these was the French mathematician, engineer, and architect Girard Desargues who, in 1536, published a geometric method for making perspective images of objects. His ideas powerfully influenced some artists of the time, including the painter Laurent de La Hyre and the engraver Abraham Bosse, but were subsequently forgotten about until the early 1800s.

Three centuries after Brunelleschi showed how 3D objects could be projected onto a flat surface perpendicular to the line of sight, the French mathematician Jean-Victor Poncelet put projective geometry literally onto a new plane. While serving in Napoleon's army in Russia, he was captured, forced to march for five months across frozen plains, and thrown into prison in Saratov on the lower Volga. During his incarceration, from March 1813 to June 1814, he wrote down some of his discoveries and later, in 1822, published them as *Traité des propriétés projectives des figures*. Effectively, he generalised Brunelleschi's findings to include planes that are inclined or rotated. Early in the twentieth century, the Dutch mathematician and philosopher L. E. J. Brouwer went further and broadened Poncelet's findings to embrace projections on surfaces that could be stretched or twisted into any shape, as if they were made of rubber. Finally, the story of projective geometry came full circle, back to art, when the American sculptor and outdoor artist Jim Sanborn brought Brouwer's principles to life, by projecting a pattern of concentric circles onto a rock formation at Kilkee on the coast of County Clare, Ireland, in 1997.

For more than 500 years, developments in this field ricocheted back and forth between mathematicians and artists. Even physicists got in on the act. The Englishman Paul Dirac, who earned a degree in mathematics before moving into theoretical physics, remarked that projective geometry was his favourite mathematical topic and that it served as a source of physical insight. Although he never said so explicitly, there's evidence to suggest it played a role in the development of his famous equation in quantum mechanics. The Dirac equation shows how particles like electrons behave when they travel close to the speed of light and also predicted the existence of antimatter.

Some well-known artists have gone out of their way to infuse mathematical ideas into their work. One of the earliest was the German painter and printmaker Albrecht Dürer, who was also a practising mathematician. In one of his greatest and most influential pieces, an exquisite copper engraving made in 1514 called *Melencolia I*, he depicts a winged figure representative of one of the four 'temperaments' of mediaeval philosophy. Each temperament matched one of four 'humours', or distinct bodily fluids. Melancholy was associated with black gall and also with the god Saturn and a tendency to creative genius and insanity. In her lap, in Dürer's depiction, Melancholy holds a book, and in her right hand a pair of dividers. Around her are various mathematical objects including a sphere, an unusual polyhedron, and a 4 by 4 magic square – an arrangement of the numbers 1 to 16 in which each number appears just once, and the sum of the numbers in every row, column, and main diagonal is the same (34 in this case). Coded in the square are the year *Melencolia I* was drawn (the numbers 15 and 14 appear in the bottom row) and Dürer's age at the time and initials. Knowledge of magic squares goes back more than 2,000 years to the Chinese, but Dürer was the first to bring them to widespread public attention and to inspire serious mathematical study of them in the West. In time, the prolific Swiss mathematician and physicist Leonhard Euler wrote a paper, 'On Magic Squares' (1776), defining what came to be known as Euler squares, which, in turn, found application in modern developments in combinatorics (the maths of combinations and permutations) and efficient radio communication by frequency-hopping.

Polyhedra were another of Dürer's fascinations. The one that appears in *Melencolia I* has attracted a lot of speculation and controversy up to the present day. It's an eight-faced solid known technically as a truncated triangular trapezohedron. It can be made by balancing a cube on one of its corners, stretching it a bit vertically between the top and bottom corners, and then slicing off these two corners horizontally. The mystery is why Dürer chose to illustrate this particular, fairly obscure shape. It's a form adopted by some crystals, such as calcite, but Dürer could hardly have known that because the mathematical study of crystals didn't start for another century or so. A couple of other possibilities arise from sketches in the artist's notebooks. One of these depicts a shape like the Melencolia solid drawn so that it fits inside a sphere, a property shared with the five famous Platonic solids. Another suggests that Dürer proportioned his shape so as to give an approximate solution to the classical problem of doubling the volume of a cube using only a compass and straightedge (an unmarked ruler). We know now that this so-called 'Delian problem' can't be solved exactly but Dürer does describe in detail an excellent approximate solution in his geometry book *Underweysung der Messung mit dem Zyrkel und Rychtscheyd* ('Instruction in measurement with compass and ruler'), published in 1525, just a few years before his death.

Melencolia I, copper engraving by Albrecht Dürer.

Dürer's book also introduces a new way of teaching geometry by folding polygons (flat, straight-sided shapes) into 3D polyhedra. Schoolchildren around the world are now familiar with this idea of making polyhedra, such as cubes or pyramids, from 'nets' – polygons that are joined along some of their edges. A familiar problem in exams is to decide which of a number of different nets shown can be folded to make a given 3D shape.

A polyhedron can be made from various different nets, depending on which of the edges are joined or separated.

It's also true that a given net may be folded into more than one different convex polyhedron, depending on which edges are fixed together and the angles at which they're folded. A polyhedron is convex if a line that connects any two points on the surface of the polyhedron lies completely inside or on the surface of the shape. In 1975, the English mathematician Geoffrey Shephard posed a question about polyhedra that remains unanswered. This open problem, which is sometimes called Dürer's conjecture, in reference to the German's pioneering work on the subject, asks whether every convex polyhedron has at least one net. There are certainly non-convex polyhedra that don't have nets. It's also known that faces of all convex polyhedra can be subdivided so that the set of subdivided faces has a net. But the general problem posed by Shephard is still unsolved.

The tradition of art illustrating maths and maths inspiring art has never been more lively than in recent times. In the twentieth century, the Dutchman Maurits Escher and the Spaniard Salvador Dalí are perhaps the best-known artists to make the portrayal of concepts derived from maths and science a central feature of their compositions. Both worked closely with prominent mathematicians and scientists in the realisation of their drawings, etchings, and paintings, and both helped provide a new way of appreciating ideas that, in their original academic form, were hard to grasp.

Escher claimed to have no mathematical talent himself but eventually came to collaborate closely with a number of leading figures in mathematics and science, including the Hungarian George Pólya, the Englishman Roger Penrose, the Canadian Harold Coxeter, and the German crystallographer Friedrich Haag. A sickly child, Escher struggled in school, but, beginning in his twenties, found artistic inspiration from his travels around Italy and Spain, especially from the exquisite decorative designs in the Alhambra, a Moorish palace and fortress in Granada. The fabulous variety and intricacy of the Alhambran tilings fired Escher's interest in tessellations (more of which in Chapter 7) and found expression in some of his most famous pieces.

Escher's absorption in mathematics, and his skilful, easily accessible portrayals of technical ideas, drew criticism from many in the art world for being overly intellectual but attracted a huge popular following that continues to this day. His works adorn countless posters and book and album covers, including Douglas Hofstadter's best-selling *Gödel, Escher, Bach* and Mott the Hoople's eponymous 1969 LP. Beyond tessellations, of animals and other figures, he explored recursion, different dimensions, and, perhaps most famously, impossible constructions – pictures that seem to make sense in places but, as a whole, boggle the mind and disorient the viewer. *Relativity* (1953) depicts a building in which gravity appears to operate in three different directions, leading to a series of baffling, inconsistent viewpoints and impossibly connected stairways. Displayed alongside some of Escher's other work at a museum in Amsterdam in 1954, to coincide with the International Congress of Mathematicians, held in the city that year, *Relativity* attracted the attention of the mathematical physicist Roger Penrose and the geometer Harold Coxeter.

Inspired by Escher's lithograph, Penrose and his father, Lionel – a psychiatrist, geneticist, and mathematician – began their own explorations of impossible objects: shapes that can be drawn in two dimensions but defy realisation in three. A few years after the Amsterdam conference, Roger sent Escher his sketch of what has become known as the Penrose tribar or Penrose triangle – a shape first drawn in 1934 by the Swedish artist Oscar Reutersvärd. He also included a picture by his father of an endless staircase. The latter, in turn, inspired Escher to create *Ascending and Descending* (1960) and *Waterfall* (1961), both of which depict something – monk-like figures and water, respectively – climbing (or descending) only to end up at their starting point.

Recognising, from the museum exhibition, Escher's interest and talent in drawing complex tessellations, Howard Coxeter sent the artist a copy of a paper he'd delivered at the Amsterdam conference. This included a diagram that Coxeter had prepared himself of a tessellated hyperbolic plane. Like the ordinary Euclidean plane with which everyone's familiar, a hyperbolic plane is open and stretches away to infinity. It differs, though, in that parallel lines on it can meet or intersect in one direction and diverge in another. If the hyperbolic plane is represented as a disc, shapes, such as triangles, drawn on it become more and more distorted and crowded together towards the disc's edge. When Escher saw Coxeter's drawing of such a disc, known as a Poincaré disc, he immediately grasped that here was a way of representing infinity on a finite 2D plane. Helped by further discussions with Coxeter, he then set about producing his own tilings of the hyperbolic plane using more complex shapes. The result was *Circle Limit I–IV* (1958–1960), a series of four wood engravings culminating in *Circle Limit IV: Heaven and Hell* in which a tessellation of white angels and black devils is mapped onto a Poincaré disc.

Salvador Dalí, a contemporary of Escher's, also worked closely with mathematicians and scientists. In 1955, seeing Dalí's *Crucifixion (Corpus Hypercubus)*, a painting of Christ pinioned to the polyhedron net of a tesseract (the

4D equivalent of a cube), in the Met, helped fire a young Thomas Banchoff's interest in higher dimensions. Twenty years later, now a professor of maths at Brown University, Providence, Rhode Island, Banchoff received an invitation from Dalí to meet him in New York. One of Banchoff's colleagues quipped: 'It's either a hoax or a lawsuit.' In fact, Dalí was embarking on a series of stereoscopic paintings and wanted help with viewing techniques. It was to be the start of a decade-long collaboration between the two men.

In the 1950s, the focus of Dalí's interest had moved from psychology to science and maths. Of this transition, he wrote:

> In the Surrealist period I wanted to create the ico-
> nography of the interior world and the world of the
> marvellous, of my father Freud ... Today the exterior
> world and that of physics, has transcended the one
> of psychology. My father today is Dr Heisenberg.

Dalí's paradigm shift is nowhere more clearly revealed than in the contrast between his paintings *The Persistence of Memory* (1931) and *The Disintegration of the Persistence of Memory* (1954). In the former, one of his most instantly recognisable works, soft pocket watches draped like cloth over various objects suggest the fluidity of time and space as experienced in dreams and other altered states of consciousness. By contrast, *Disintegration*, with its blocks and fragmentation of the older scene, takes the viewpoint of the modern physicist in which matter and energy are broken down into discrete quanta.

In *The Sacrament of the Last Supper* (1955), one of the most popular paintings from his postwar period of fascination with science, religion, and geometry, Dalí incorporates the famous – not to say infamous – golden ratio. No other quantity in maths has proved more alluring to artists, scientists, psychologists, numerologists, and, in some cases, mathematicians themselves, yet been so misrepresented. The golden ratio is both curious and important, but it's also been the source of many false claims.

Two quantities, a larger one a and a smaller one b, are said to be in the golden ratio if their ratio, a/b, is equal to the ratio of their sum to the larger quantity, $(a + b)/a$. The value of the golden ratio, represented by the Greek letter phi (ϕ) is $(1 + \sqrt{5})/2$, or 1.6180339887… Like pi, it's irrational, in other words it can't be written as one whole number divided by another whole number, and so its decimal expansion goes on forever without having a recurring pattern. But unlike pi, it *can* be written as the solution of an equation in algebra with integer coefficients (the '5' in $5x^2$, for example, is a coefficient), which means that it isn't transcendental.

If a rectangle is drawn with its sides in the golden ratio, it's called, not surprisingly, a golden rectangle. This is the shape chosen by Dalí for his *Sacrament*, the canvas dimensions of which are 166.7 by 267 centimetres (65 5/8 by 105 1/8 inches). He also placed the top of the table at the golden ratio of the height of the painting, and the two disciples immediately to the right and left of Christ at the golden ratios of the width of the piece. The scene is set within a large dodecahedron, a 12-sided shape with pentagonal windows that look out on a scene revealing the landscape of Dalí's native Catalonia. Forming the centre of each pentagonal face of a regular dodecahedron is the intersection of three golden rectangles, while rectangles of the ratio $(\phi + 1):1$ and $\phi:1$ also fit precisely within a regular dodecahedron.

Dalí may have borrowed from Leonardo da Vinci in using phi in his depiction of this Biblical event. Certain dimensions of the room, the table, and other elements of Leonardo's *The Last Supper* appear to conform to the golden ratio, though whether some do so just coincidentally is impossible to say. It's also been pointed out that a golden rectangle can be drawn around the face of Leonardo's most famous work, the *Mona Lisa*. Again, we can't be sure if this was intentional and, in any case, it's hard to be precise about where the framing rectangle should be drawn. What's beyond doubt, however, is that Leonardo was a close friend of the mathematician and Franciscan friar Luca Pacioli, who published a three-volume treatise on the golden ratio in 1509 called *De Divina Proportione*, which Leonardo illustrated. That name – the 'divine proportion' – was used by many Renaissance thinkers in referring to phi and reflects the mystical reverence with which the number was often held.

Phi is, indeed, a remarkable quantity in maths and, like pi, crops up in all sorts of unexpected places. It's intimately linked, for instance, to the Fibonacci sequence, first described by Leonardo Fibonacci around 1200. This sequence starts with 0 and 1 and then proceeds by simply adding the previous two members together: 0, 1, 1, 2, 3, 5, 8, 13, and so on. The ratio of two successive Fibonacci numbers approaches the value of phi as the numbers get larger: $3/2 = 1.5$, $13/8 = 1.625$, $233/144 = 1.618$, and so on. Drawing a curve within adjacent rectangles of Fibonacci dimensions produces a spiral pattern that is often seen in nature – in the shape of shells and waves, and in the arrangement of seeds on a sunflower or petals on a rose. The close connection between the golden ratio and the Fibonacci series ensures that a similar intimacy exists between the natural world and the 'golden spiral', generated from adjacent golden rectangles, of which, as we've seen, Fibonacci rectangles are an approximation.

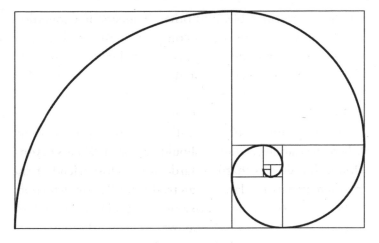

Fibonacci spiral.

Given this ubiquity of phi in maths and its tendency to show up in the most unexpected places in nature it's hardly surprising that Renaissance thinkers granted it 'divine' status. This was a time – from the fourteenth to the seventeenth century – when intellectuals were striving to unify their fast expanding cosmos of ideas, including things both on Earth and beyond, within a philosophy that allowed for supernatural intervention. Among the scientists at the heart of this effort was the German astronomer and mathematician Johannes Kepler, who was obsessed with the idea that the universe was organised according to strict rules of harmony, balance, and mathematical symmetry. In his *Harmonices Mundi* ('Harmony of the Worlds'), he wrote about the eternal geometric forms that guided and spaced the planets, and the musical sounds that the objects in the heavens made as they trekked on their eternal ways. In his essay *De nive sexangula* ('On the Six Cornered Snowflake'), he discoursed on the divine proportion (phi) and the Fibonacci sequence alongside regular polygons and flower petals. 'Geometry', he wrote, 'has two great treasures; one is the Theorem of Pythagoras; the other, the division of a line into extreme and mean ratio. The first we may compare to a measure of gold, the second we may name a precious jewel.'

Phi may have a habit of insinuating itself into some of the most curious and surprising areas of maths, but some claims about it have been overstated or are just plain wrong. Numerologists and pseudo-historians love to seek out relationships where none, in all likelihood, exist. The ratio of the base to the height of the Great Pyramid of Giza, for example, does not equal phi, as pyramidologists would have us believe. Instead, it equals 1.572. Nor does the golden ratio describe the shape of another great structure from classical times – the Parthenon in Athens.

In more recent times, some researchers claim to have found scientific evidence that the golden ratio holds a uniquely aesthetic appeal for the human mind and senses. The German physicist and psychologist Gustav Fechner started off this trend with a series of experiments in the 1860s involving a variety of rectangles with different length-to-width ratios. Subjects were invited to pick which they found most appealing in shape. He found that three quarters of all choices involved just three of the rectangles, with ratios of 1.50, 1.62, and 1.75, and that the most popular centred on the golden ratio at 1.62. He then went on to measure the ratios of thousands of different rectangular-shaped objects, including window frames, picture frames in museums, and books in a library. The average ratio, he claimed in his book *Vorschule der Aesthetik* ('Introduction to Aesthetics') was very close to the golden ratio.

Fechner's discovery, however, has been disputed. The Canadian psychologist Michael Godkewitsch has argued that his conclusion is flawed because a subject's choice of preferred rectangle may vary depending on the position of the rectangle in the range on offer. Similar scepticism was expressed by the British psychologist Chris McManus who commented: 'Whether the golden section [another name for the golden ratio] *per se* is important, as opposed to similar ratios (e.g. 1.5, 1.6 or even 1.75), is very unclear.'

The issue surfaced again with the suggestion, by some, that people's faces are likely to be considered more attractive if certain proportions of them follow the golden ratio. Orthodontist Mark Lowey, at University College Hospital, London, published results in 1994 suggesting that the faces of fashion models were regarded as beautiful because they showed this tendency. But this, too, has been disputed. In a different study at the Maxillofacial Unit of the same hospital, Alfred Linney and his colleagues used lasers to make precise measurements of the faces of top models. They found that the facial features of the models were just as varied as those in the rest of the population.

By their very nature, art and architecture have strongly subjective elements. In their appeal to the human senses and emotions, they venture into areas of experience untouched by maths in its highest, purest form. Although this may result in some loss of precision, or the infusion of features that are extraneous to maths, it offers a way to appreciate mathematical beauty without the need for great intellectual effort or years of study. The Australian mathematician Henry Segerman, who makes 3D printed models to help his students better understand mathematical formulae, put it this way: 'The language of mathematics is often less accessible than the language of art, but I can try to translate from one to the other, producing a picture or sculpture that expresses a mathematical idea.'

Today, equipped with the tools of the digital revolution, creative individuals can bring to life concepts in advanced mathematics that only a handful of specialists can truly grasp. The American sculptor Jim Sanborn, mentioned earlier, specialises in 'making the invisible visible', through pieces that address topics such as magnetism, nuclear reactions, and cryptography. His sculpture *Kryptos*, which sits outside the headquarters of the Central Intelligence Agency in Langley, Virginia, contains four messages encoded in its 2,000 alphabetic characters, one of which remains undeciphered. His *Coastline*, installed at the National Oceanic and Atmospheric Administration's complex in Silver Spring, Maryland, includes a turbine and pneumatic blower that recreates in real time, on a miniature scale, the waves breaking near the shore at NOAA's monitoring station at Woods Hole, Massachusetts, on the Atlantic coast.

Fractals – objects that show structure and complexity at all scales – can result in gorgeous, mesmerising patterns that are an artist's dream. British laser physicist-turned-artist Tom Beddard creates 3D digital renderings of Fabergé eggs adorned with fractal patterns. He's one of many creative individuals who are harnessing the power of computers to reveal the hidden beauty in formulae that otherwise, to most people, would be meaningless strings of symbols and operations. 'The formulae effectively fold, scale, rotate, or flip space,' said Beddard. But in the absence of artistic representation, the spectacle of these technical gymnastics would go unappreciated.

In a sense, art and maths represent opposite ends of the spectrum of experience. Art the subjective, the passionate, the sensual; maths, the ruthlessly logical and cerebral. And between the two, the point of connection, none other than ourselves – conscious observers of a reality endlessly rich in possibility.

CHAPTER 6

Beyond the Imaginary

> Imaginary numbers are a fine and wonderful refuge of
> the divine spirit – almost an amphibian between being
> and non-being.
>
> – Gottfried Leibniz

IN A REMOTE Amazonian region of Brazil lives a tribe, the Pirahã, whose couple of hundred members can't count beyond two. Their word for 'one' can also mean 'a few', while 'two' does double duty as 'not many'. Anything else is simply 'many'. They also have no way of saying 'more', 'several', or 'all'. To us, this seems very strange. After all, it's normal for toddlers who've just turned two to be able to count to three, and a year later to be able to count to five or so. But the Pirahã aren't stupid. They're hunter-gatherers who have no need to count and so no need to practise doing it. American linguist Daniel Everett tried to teach the Pirahã some basic numeracy skills after they expressed concern that their lack of knowledge might make it easy for them to be cheated when trading with other tribes. After eight months' effort, however, not a single Pirahã had learned how to count to 10 or even to add one and one. Both their culture and their previous experience left them totally unprepared to grasp even the rudiments of numbers.

We're so used to numbers from an early age that we forget that there's nothing obvious about them. They're not like things, animals, and people in the everyday world that parents can point out to their children and label 'flower', 'dog', or 'eye'. They're abstractions and, as the Pirahã example shows, ones that are hard to grasp unless we're exposed to them from an early age – which most of us are. Even so, some numbers are easier to understand than others. A three-year old, for example, if encouraged, might be able to count to 10 or more, but probably doesn't have a real understanding of the meaning of numbers much larger than 3. Adding comes a little later, fractions and how to deal with them follow, and finally we're introduced to the mysteries of working with negative numbers. None of these types of numbers are self-evident. Much less so are numbers that we never use in our day-to-day lives and that many of us never study in school – so-called imaginary numbers and, beyond them, exotic beasts such as surreals and transfinites. Yet, in maths, all these dwellers of the numerical cosmos are equally 'real' and valid, even though they may seem as unintelligible and irrelevant to most of us as 3, 4, and 5 are to an Amazonian hunter-gatherer.

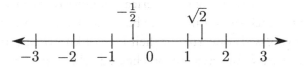

The number line.

Quite early on in our schooling, we're introduced to the idea of the number line, starting from 0 and extending in one direction towards bigger and bigger values. Then, negative numbers are brought into the picture and we learn that the number line also extends the other way, as far as we care to go. The integers, positive and negative, and zero, quickly become concepts with which we're familiar and comfortable. How could they not be clear to everyone? Yet for much of human history, the number line would have seemed completely alien.

We don't know when numbers were first used. Some animals, including birds and rodents, can tell at a glance when one pile of objects is bigger than another. There's an obvious survival advantage in being able to do this. But it isn't the same as counting. To count, an animal would have to recognise, at some level, that each object in a collection corresponds to a single number and that the last number in a sequence represents the total number of objects. Research has shown that not only many primates have this innate ability, but so too do dogs. Animal behaviour researchers Robert Young and Rebecca West carried out an experiment, in 2002, using eleven mongrels and some dog treats. A few treats were put in a bowl in front of each dog, and then a screen was raised to block the dog's view. The dog then watched as some treats were taken away or added before the screen was lowered again. If the researcher surreptitiously took away or added more treats than were shown, the dog would stare at the bowl much longer, apparently aware that the sums didn't add up.

Numerals – symbols for numbers – and rules for doing simple arithmetic followed the rise of the first civilisations, in Sumer and other parts of Mesopotamia. But there's good evidence that people kept track of the number of things (though exactly *what* things isn't clear) much earlier, in the form of tally sticks. The Lebombo bone found in Border Cave in the Lebombo Mountains, bordering Swaziland and South Africa, is a baboon fibula, at least 43,000 years old, on which 29 notches have been made. One theory is that it was used to keep track of the phases of the Moon, in which case African women may have been the first mathematicians, because menstrual cycles are linked to the lunar calendar. Others dispute this, however, pointing out that the bone is broken and may originally have had more than 29 marks. It's also been suggested that the markings are purely decorative. More complex are the markings on another bone tool, the Ishango bone, found in 1960 near the Semliki River on the Uganda–Congo border, and dating back perhaps 20,000 years or more. The exact interpretation of the Ishango notches is again a matter of debate but some of the patterns hint at a surprisingly sophisticated knowledge of maths that would long predate civilisation. The bone (another baboon fibula) has a series of notches carved in groups of three rows running the length of the bone. The markings on two of these rows each add to 60. The first row is consistent with a number system based on 10, since the notches are grouped as 20 + 1,

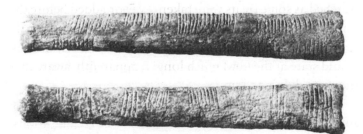

The Ishango bone.

20 – 1, 10 + 1, and 10 – 1, while the second row contains the prime numbers between 10 and 20. A third seems to show a method for multiplying by 2 that was employed much later by the Egyptians. Whether these are mere coincidences we can't be certain. However, a second bone, found a year earlier, also bears a pattern of notches suggestive of an understanding of number and number bases.

What we know for sure is that when people began to settle in towns and cities in the Middle East, several thousand years BCE, they had a need for numbers and began to develop ways to represent them and do basic operations with them, such as addition and subtraction. The need arose from trade and the importance of keeping accurate track of transactions. For example, if I agreed to give you ten sheep as part of a deal, you had to be sure that I wasn't conning you by handing over only nine sheep! A reliable method of counting became essential because most of us can't immediately judge the difference between nine things and ten. Only through awareness and use of the natural numbers 1, 2, 3, 4 … is this possible. At this stage, however, no one gave a thought as to what might come between or before these numbers.

Earlier, before commerce and business appeared on the scene, natural numbers weren't a necessity. If I'm a shepherd with 10 or 20 sheep, it's not essential that I know the exact number: an approximate idea is good enough. It's only with the rise in importance of trade that the naturals became an indispensable part of our lives. To begin with, they took the form of sealed clay tokens called *bullae*, but later a system of writing down numbers in a way similar to tally marks evolved. At this stage, people still hadn't grasped the idea that numbers could be distinct from what was being counted, so that at first the number ten, say, wasn't treated as an entity in itself, common to ten sheep, ten cows, or ten loaves of bread. The notion of natural numbers as having a separate existence from the collections of things they describe took time to develop. But, when it did, it had a powerful effect on mathematics and how we think.

Eventually, as city-states formed and not everyone had to be busy doing practical chores all day long to make a living, philosophers and others who simply thought and taught about the world entered the scene. In Greece, Pythagoras and his followers rose to prominence in the sixth century BCE propagating the belief that natural numbers were the key to the universe – that all in essence stemmed from these timeless, perfect, abstract creations that stood behind the reality we see. Each whole number, the Pythagoreans believed, represented something different, and the relationships between natural numbers generated everything else. Another celebrated Greek, Euclid, in his magnum opus, *Elements*, came up with many theorems about the natural numbers in addition to his work on geometry, the best known being his proof that there are infinitely many primes. Yet it took until the seventh century CE before we broke out of the confines of the numbers with which every child today learns to count.

An Indian mathematician, Brahmagupta, was the first we know of to go beyond the naturals – and he did it in two different ways at once. He described rules for arithmetic that deal not only with zero but also with negative numbers. It's very likely that people before him had some inkling about how to work with these new numbers, but he's the first of whom we have a clear record. Adding zero to the natural numbers results in the whole numbers and has an importance that goes beyond zero as a value (as we saw in Chapter 2). But adding the negative numbers is a much greater extension because it means that now there's no beginning to the number system: the number line extends to infinity in both directions.

Left to their own devices, merchants, farmers, or anyone else using maths just to do simple reckoning would probably never have thought of the idea for zero or negative numbers. Who ever heard of having minus six horses, for instance? Negative things don't exist in daily life. And why bother to have zero as a number when it doesn't change anything if you add it or take it away? It took philosophers and theoreticians – abstract thinkers – to come up with these strange possibilities and, in so doing, broaden our mathematical horizons. Brahmagupta did point out, though, that negative numbers have a very practical use: as a way of representing debt. If you owe someone three cows and have none, you effectively have minus 3 cows!

Today, negative numbers don't seem strange because we're taught about them early on when our brains can adapt effortlessly to them. Besides, we're used to thermometers reading 'below zero' when it's very cold. But even into the Renaissance, negative numbers were hugely controversial in the mathematical world. When a solution to a problem was a negative number, it would often be described as 'fictitious'. Only slowly did anything left of zero on the number line gain respectability.

Mathematicians, going back at least as far as Pythagoras, were more accepting of another type of number that extended the naturals. Of course, the Pythagoreans loved natural numbers. Nothing, in their eyes, could match the perfection of 1, 2, 3, 4 ..., or their importance in underpinning the whole cosmos. But they were content to allow the existence of *rational* numbers – the result of dividing one whole number by another. Pythagoras saw rationals in terms of a relationship between two natural numbers rather than being numbers in their own right, but that didn't stop him from using them in his mathematics. He and his followers believed that all numbers could be expressed as ratios. But they were wrong – perhaps tragically so.

There are many slightly whacky tales about Pythagoras, most of them doubtless apocryphal, and this may be one of them. But there's a story that one of his students had made the shocking discovery that the square root of 2 – the length of the hypotenuse of a right-angled triangle whose shorter sides are both 1 in length – couldn't be expressed as any whole-number ratio at all. For this unspeakable sin, Hippasus was, if the tale's to be believed, drowned, either by the great mathematician himself or one or more of his band of ardent supporters.

No amount of irrational behaviour, though, could deny the fact that irrational numbers – numbers that can't be expressed as fractions – really did exist. In time they took their place alongside the rationals in making up the complete number line. Rational numbers and irrationals together make up the so-called real numbers. Mathematicians came to accept the reality of the reals and knew what they were. The thing they couldn't do, not for a very long time, was give a formal definition of them. The natural numbers were easy to define and generate – you could represent them using 1 and the successor operation, which simply adds 1 to a natural number. This allows all natural numbers to be represented. The integers were a simple extension once the natural numbers had been defined – just include 0 as well as the negatives of the natural numbers. Rationals, too, were easy to produce as they came from just dividing two integers (providing the denominator wasn't zero). But how could we use the rational numbers as a springboard for reaching the reals? The problem was finally solved, as late as the nineteenth century, by the German mathematician Richard Dedekind.

Dedekind used what are now known as Dedekind cuts to define real numbers. A Dedekind cut simply divides the set of rational numbers into two sets in such a way that every number in the first set is smaller than any in the second. For example, one Dedekind cut might be to divide the rationals into the following sets: first, those rational numbers x that are negative or satisfy the condition that x^2 must be less than 2, and secondly, those x that are positive and for which x^2 is greater than 2. So, for instance, 1, 1.4, 1.41, 1.414, and 1.4142 are all members of the first set, and 2, 1.5, 1.42, 1.415, and 1.4143 are all members of the second set. This Dedekind cut, using sets of only rationals, defines the real number $\sqrt{2}$, which is irrational. (The restriction that any negative x belongs in the first set is required to prevent negative numbers with a square greater than 2, such as -2, from being in the second set.) The cut is based on the idea of approximating a real number more and more closely with decimals (or any such notation) in a way that can be formalised – and it can be used to generate any irrational from two sets of rationals.

So, we have the real number line and the knowledge that, if we care to, we can formally define any number that lies on it. The term 'real' suggests that this is the end of story as far as numbers that matter are concerned. Science fiction writers might be interested in stories about numbers that aren't real – ones that populate fantastic universes governed by different rules of logic. But surely there's no place in maths for 'unreal' numbers. The problem is that the names, which have arisen historically for different types of numbers, are totally misleading. Real numbers that aren't rational are said to be irrational. The first definition in the Oxford English Dictionary of 'irrational' is 'not logical or reasonable'. Only further down the page do we find the special meaning in maths: '(of a number, quantity, or expression) not express-ible as a ratio of two integers'. As for 'real', the OED has this as its leading definition: 'Actually existing as a thing or occurring in fact; not imagined or supposed.'

Surely no mathematician worth their salt would be inter-ested in a number that was 'imagined or supposed'. That was certainly the attitude of many mathematicians up until the eighteenth century. Any suggestion that there might exist numbers that didn't lie on the real number line was treated as tantamount to witchcraft. But there was the sticky prob-lem of what to do about things such as $\sqrt{-2}$. The square root of plain old 2 was controversial enough back in the day, as rumours about Hippasus and his untimely demise attest. But the square root of *minus* 2 – what's that all about? There's no such animal among the real numbers, that's for sure. The only option mathematicians had was to ignore it, denounce it as being 'fictitious' (as in the case of negative numbers), and hope that it went away, or embrace it and welcome it into the mathematical fold.

The first to entertain the actual existence of numbers that allowed for taking the square root of negative real numbers *and* who set out rules for dealing with these novelties was Italian mathematician Rafael Bombelli. In his *L'Algebra*, published in 1572, Bombelli became the first European clearly to state ways of doing arithmetic with negative numbers that made sense (such as 'minus times plus equals minus'). But more importantly he launched the study of what even-tually became known as complex numbers by considering the solutions of equations such as $x^3 = ax + b$ in situations where $(a/x)^3$ is bigger than $(b/x)^2$. The only way to crack the equation in such a case is to allow the existence of something that's the sum of a real number plus the square root of a negative real number.

For well over a century, mention of 'square roots of negative real numbers' met with little enthusiasm among the mathematical cognoscenti. Bombelli was smart enough not to give such things a special name, and thereby expose them to even more ridicule. But it wasn't long before 'imaginary numbers' became the term of derision used in an attempt to discredit the whole idea. Unfortunately, the name stuck and so, even today, in these more enlightened times, we refer to $\sqrt{-1}$ as the unit imaginary number and represent it by the letter i. Any real number multiples of i, such as $5i$, πi, or $i\sqrt{2}$ (which is equal to $\sqrt{-2}$) are known as imaginary numbers – even though they're as real as real numbers! The sum of a real number and an imaginary number is known as a complex number. Again 'complex' is a misnomer because it implies complexity in the everyday sense of being difficult or complicated, which isn't the case at all. Many people never study complex numbers in school but one of us (David) often introduces the idea of imaginary and complex numbers to his private students as young as ten or eleven and they have no trouble grasping it.

Historically, complex numbers took off and, in time, became accepted because they proved useful as intermediate steps in obtaining real-number answers to problems. It's true that we don't need them in everyday maths. We can even get by, for the most part, without having to know about negative numbers. Why would we need to worry about dealing with an i number of things? Hardly any of us use imaginary numbers on a day-to-day basis but all of us depend on others who do know about and use them because these numbers are crucial in many fields of modern physics and engineering. They're used by electrical engineers as a method of representing alternating current and are unavoidable in areas of physics such as the theory of relativity and quantum mechanics (which underpins our understanding of the world at the atomic and subatomic level). This relevance in science arises because complex numbers have some very useful mathematical properties. For instance, polynomials, such as $x^2 + 1 = 0$, may not always have a solution among the real numbers but will *always* have a solution in the complex numbers. This fact was first proven by German mathematician Carl Gauss in 1799 and is so important it's known as the Fundamental Theorem of Algebra.

Surely, with complex numbers, we've reached the end of the road in terms of what's mathematically possible. But no, far from it. Larger even than the system of complex numbers are other systems so vast that the only way to understand them is to venture into a strange land known as abstract algebra. This esoteric realm of maths – especially appealing to those who enjoy building their own private thought universes – describes, in the broadest sense, sets (collections of things) upon which certain well-defined operations can be performed. One type of object studied in abstract algebra is the group, some examples of which we met in Chapter 4 when exploring symmetry. Another is the ring, which has nothing to do with circles but is instead a set on which two operations are defined, labelled + and ×. These operations share the same properties as the addition and multiplication with which we're familiar. To be precise, when it comes to rings, addition must be associative, and there must be both an identity and an additive inverse. The multiplication of rings also has to satisfy several conditions. The natural numbers don't form a ring because there's no additive identity or additive inverse (0 isn't a natural number, nor are negative numbers). But the integers *do* form a ring. Other rings include the rationals, the reals, and the complex numbers, and there are many other examples.

Abstract algebra can help us to define new systems of numbers and classify them based on whether they're rings or some other type of mathematical object. We can find rings that aren't simple extensions of the integers and we can also find much larger systems of numbers. One of these is the quaternions, discovered in 1843 by Irish mathematician William Hamilton. Complex numbers can be represented on a two-dimensional plane on which the x-axis represents real numbers and the y-axis imaginary numbers. Hamilton wondered whether a larger system than complex numbers could be represented in 3D space. He struggled to find one but eventually hit upon quaternions, which he pictured as existing in a space of four dimensions. In a moment of inspiration, while strolling across Brougham Bridge in Dublin, the formula $i^2 = j^2 = k^2 = ijk = -1$ flashed into his mind, along with the realisation that there weren't just two possible square roots of -1 (i and $-i$) but six. In fact we now know that there are infinitely many square roots of -1!

Quaternions never caught on widely but have proved their worth for some applications, even though they remain more obscure to most people even than complex numbers. In cases where a quaternion consists only of a multiple of i, j, and k, it corresponds to a vector (a quantity with size and direction) in 3D space. In fact, a quaternion can be represented as the sum of a vector and a scalar, with the scalar being the real-number part. This way of representing

3D vectors makes quaternions extremely useful in three-dimensional animations and simulations, where the ability to rotate a perspective is essential. Computer games that have a three-dimensional display, for example, use quaternions to represent such rotations.

Inspired by Hamilton's discovery, fellow Irish mathematician John Graves came up with yet another new system of numbers, which he called octonions. He was slow to publish his findings, however, and was pipped to the post by Englishman Arthur Cayley who introduced octonions to the world in 1845. Octonions are sums of multiples of 1 and seven other values (often simply called e_1, e_2 ... e_7). They satisfy the equality $e_1^2 = e_2^2 = ... = e_7^2 = -1$. But for multiplying two distinct octonions, a much more complicated multiplication table is required. Obscure though they are, octonions have found some applications at the cutting edge of physics in the highly mathematical subject of string theory.

Even now, we haven't reached the limits of what's possible with number systems – or the imaginations of mathematicians. Ways have been found to extend the real number line to include both the infinitely large and the infinitely small. In the system of what are called hyperreal numbers, an infinitely large number ω (omega) and an infinitely small number, or infinitesimal, ε (epsilon), are added to the real numbers. These are related by the fact that $\varepsilon = 1/\omega$. Multiples of ω and ε are allowed as well, so that, for example, $3\omega + \pi - \varepsilon\sqrt{2}$ is a hyperreal number. There are hyperreals such as ω^2, which is greater than any real multiple of ω, and ε^2, which is smaller than any real multiple of ε. Because it's possible to add, subtract, multiply, and divide them, hyperreals form a field, in the same way that rational and real numbers do. They can also be ordered, because we can define what it means for a hyperreal to be greater than another, so they're known as an ordered field. Some other fields, like the field of complex numbers, aren't ordered. How, for example, can we tell whether i is greater than or less than 0 when it lies to the side of our number line?

The richest extension of the real numbers, which includes an infinity of infinities and infinitesimals, are the surreal numbers, which we met earlier. The surreals take the concept of Dedekind cuts to its logical extreme. They're represented in the form $\{L \mid R\}$ where L and R are sets of surreal numbers previously constructed in such a way that all members in L are less than any in R. The new surreal number then has to lie between these sets, being greater than all members of L and less than all members of R.

We can generate the whole, inconceivably vast universe of the surreals effectively out of thin air. To construct the first one, both L and R must be the empty set (the set that contains no members). This gives us the surreal number $\{\mid\}$, which is 0. With 0 in place, it can be used in the sets L and R to produce more surreals. The next two to be constructed are -1, which is $\{\mid 0\}$ and 1, which is $\{0 \mid\}$. Moving on, 2 is $\{1 \mid\}$, 3 is $\{2 \mid\}$, and so on, while $\{0 \mid 1\}$ is 1/2). All fractions where the denominator is a power of 2, known as dyadic rationals, can be expressed in the surreals in a finite number of steps. But a system that includes just dyadic rationals isn't very powerful: it can't even express all rationals, let alone all reals. The big breakthrough happens when we allow an infinite number of steps. After infinitely many steps, once all dyadic rationals have been constructed, one extra step is enough to create all the real numbers. It turns out that, while Dedekind originally used all rational numbers in his Dedekind cuts, it's enough to use just the dyadic rationals.

The reals, however, aren't the only new surreal numbers to be created. In fact, ε and ω are also made at the same time. In the case of ε, L contains 0 and R contains all previously created positive surreals (all the dyadic rationals). Meanwhile, for ω, L contains all already-existing surreals, so ω is larger than all of them. $-\varepsilon$ and $-\omega$ are also defined, as are $x + \varepsilon$ and $x - \varepsilon$ for all dyadic rationals x.

Later on, some other surreal numbers can be constructed. Once we have ω, we can have $\omega - 1$ and $\omega + 1$, and also a whole host of other numbers such as $\pi + \varepsilon$ (L consists of π, R consists of $\pi + 1$, $\pi + 1/2$, $\pi + 1/4$, and so on) and some like $\sqrt{\omega}$, where L consists of 1, 2, 3, and so on while R contains ω, $\omega/2$, $\omega/3$, $\omega/4$, ...

There are so many surreals that the real numbers amount to just an insignificant portion of them. In the same way, transcendental numbers are hugely more numerous than any of the other reals, though the disproportionality is unimaginably greater in the case of the surreals and reals.

Surreal numbers are the largest possible ordered field. They ultimately contain not only all reals but all hyperreals and even a vast hierarchy of ever-greater infinities. So mind-bogglingly huge is the number of surreals that there's no infinity large enough to contain all of them. There are so many surreal numbers that they form a proper class – there's no set large enough to contain them all.

CHAPTER 7

Tilings: Plain, Fancy, and Downright Peculiar

> At moments of great enthusiasm it seems to me that no
> one in the world has ever made something this beautiful
> and important.
>
> – M. C. Escher

ONE DAY IN San Diego in 1975, Marjorie Rice read an article in her son's copy of *Scientific American*. It stated that there were only eight known pentagonal shapes that could entirely tile, or tessellate, a plane. Despite having no maths beyond high school, she set out to find another. A couple of years later she'd come up with not just one but four new tessellations – a result noteworthy enough to be published in an academic journal.

You don't have to be a mathematician to appreciate tilings. They're almost as old as civilisation and as much works of art as they are products of intellect and reason. Their nature is simplicity itself: a tiling is a pattern made from shapes that fit together perfectly without gaps or overlaps and that can be repeated indefinitely. The tiles may be made of ceramic, brick, or other materials, and the patterns appear as deco-ration on the walls, floors, and ceilings of buildings going back to the days of ancient Sumeria.

The words 'tiling' and 'tessellation' are used interchange-ably. 'Tessellation' comes from the Latin *tessellatus* meaning 'of small *square* stones or tiles', but is used today to describe a perfectly fitting pattern made of any shaped tiles. In many tessellations, the tiles consist of regular polygons – straight-sided shapes in which all the angles are equal and all the sides are the same length. A regular tiling is made from just one kind of regular polygon. It turns out that only three types are possible: those formed from equilateral triangles, squares, or regular hexagons. The reason these three, and only these three, work is that their interior angles – 60°, 90°, and 120°, respectively – all divide into 360° exactly, which is the angle the tiles must make where their vertices, or corners, meet. A semi-regular tiling is also built up from regular polygons, but of more than one variety and in such a way that the

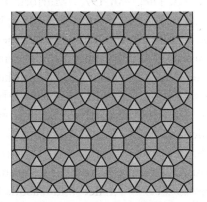

A semi-regular tiling – one that contains more than one different type of regular polygon but in which the same polygons appear around each vertex in the same order.

arrangement of polygons at every vertex is identical. There are eight of these in all, or nine if you include the tilings of equilateral triangles and hexagons that are mirror images of each other. Among the other semi-regular tilings are two that involve squares and triangles and one made from dodecagons (12-sided shapes), squares, and hexagons. Irregular tilings include every other possibility. In other words, they can be made from tiles of any shape, not just regular polygons or even shapes that have straight sides.

In nature, the most familiar example of a tessellation is the honeycomb with its array of neatly stacked hexagons. Larger hexagonal tessellations seen in formations of columnar basalt, where in the past lava has slowly cooled, occur in many places around the world, including the Giant's Causeway in Northern Ireland and the Devil's Postpile in California. Tessellated patterns are found on certain flowers, such as the fritillary, and in the scales of fish and snakes.

The known history of human-made tessellations begins around 3000 BCE, or perhaps a little earlier, with some mosaics on the columns of a Sumerian building in what is now southern Iraq. Small hexagonal tiles of different colours have been arranged to make zigzag and diamond-shaped patterns. Because the tiles are regular hexagons fitted closely together they do make genuine tessellations. This isn't the case with most mosaics, such as those often found in Roman villas, which depict scenes involving people or animals. The pieces in many mosaics, although closely spaced, have gaps between them and so fail the mathematical definition of a tiling.

In the Islamic world the representation of living things or real objects of any kind is forbidden, as it's interpreted as a kind of idolatry. Decorations in buildings were therefore restricted to purely geometrical forms. Islamic artists made the most of the limited scope in which they had to work by devising intricate and ornate patterns of shapes that locked together perfectly. Nowhere is this ingenuity better illustrated than in the fabulous palace of Alhambra, in southern Spain. Originally constructed as a small fortress in 889 CE, it was rebuilt and expanded, and finally, in the fourteenth century, converted to a magnificent place of royal residence. On the walls of the Alhambra is a tour de force of the art of tiling, breathtaking in its variety and skill.

The tile patterns at the Alhambra include arrangements not just of polygons but also of curved shapes, and tiles of different colours, put together in a celebration of both technical and aesthetic artistry. Related to the concept of tilings is that of something called wallpaper groups, of which there are 17. A wallpaper group is a mathematical way of classifying a two-dimensional repetitive pattern on the basis

A tiling at the Alhambra.

of the symmetries that the pattern displays. As we saw in Chapter 4, there are just four basic symmetry operations in 2D: reflection, rotation, translation, and glide symmetry. Every wallpaper group contains two distinct translations, so that any tiling that belongs to it can repeat endlessly and periodically to cover the entire plane. In addition, it may have other types of symmetry, including a centre of rotation, an axis of reflection, and an axis of glide reflection. It's been widely claimed that the many different tilings found at the Alhambra include representatives from all 17 wallpaper groups, although some mathematicians dispute this and say that a few of the groups may be missing. Nevertheless the variety of tilings on display is mightily impressive. Certainly it impressed and entranced the Dutch artist, Maurits Escher, who first visited the Moorish palace as a young man in 1922 before returning for a lengthier stay in 1936. He spent days sketching the tilings and taking notes, until he became completely obsessed with the idea of tessellations. Afterwards he wrote:

> It remains an extremely absorbing activity, a real mania to which I have become addicted, and from which I sometimes find it hard to tear myself away.

The sketches he made at the Alhambra became a major source of inspiration for his subsequent artwork. He delved into the maths behind what he called the 'regular division of the plane' by reading papers by the Hungarian George Pólya and the German crystallographer Friedrich Haag on plane symmetry. These papers were sent to him by his brother Berend, who was a geologist and keenly aware of the importance of symmetry in crystal structures. Escher familiarised himself with the 17 wallpaper groups and started to create periodic tilings of his own using geometric grids. In place of polygonal elements, however, he experimented with complex interlocking shapes in the form of birds, fish, reptiles, and a particularly ingenious combination of angels and devils. One of his earliest works based on tessellation and a hexagonal grid was *Study of Regular Division of the Plane with Reptiles* (1939), rendered in pencil, ink, and watercolour. The heads of three lizards, green, red, and white, meet at each vertex, while the rest of their bodies fit together precisely, leaving no gaps. It was a design that he used again in his famous lithograph *Reptiles* four years later.

The mathematical exploration of tilings, as distinct from their purely artistic expression, began only a few centuries ago. One of the first to take up the challenge was German astronomer and mathematician Johannes Kepler who wrote about tessellations in his great work *Harmonices Mundi*, published in 1619. In the first two chapters of this he tackled regular and semi-regular polygons, which led him to consider how regular and semi-regular tilings can fill the plane.

It might seem surprising that Kepler, best known for his three laws of planetary motion and who, in *Harmonices Mundi*, was mainly concerned with what he perceived as links between music theory and the movements of worlds, should include a discussion of tilings. But this was a time when mysticism and science were still entwined, and, in Kepler's mind, the perfection of the heavens must be reflected in the perfection of certain geometric forms and consonant notes on the musical scale. He was the first to investigate the mathematical structure of honeycombs and snowflakes, and the first to identify the eight forms of semi-regular tiling in addition to the three regular tilings. The former he referred to as 'perfect congruences' and the latter as 'most perfect congruences'.

Sadly, his work on tilings was largely ignored by genera-tions of mathematicians who followed and was overshadowed to a large extent by his famous astronomical work. It wasn't until the late nineteenth century that any significant further developments in the subject took place. When they did, it was in response to an urgent scientific problem: the need to classify all the various forms that crystals could take. In fact the next great leap in the maths of tilings was made by Russian Evgraf Fedorov, who combined deep interests in crystallography and geometry. Early on he was intrigued by polytopes – objects with flat sides that may exist in any number of dimensions. In 1891, six years after he published a book on this subject, *Basics of Polytopes*, he proved the two results for which he is best known. First he showed that there are exactly 230 space groups. These are all the possible symmetry groups of objects in three dimensions and represent the unique ways, in terms of their symme-try properties, that, for instance, atoms can be arranged to form crystals. On the back of this discovery, he was able to show that in two dimensions the 230 space groups reduce to just 17 types – the wallpaper groups, which we mentioned earlier.

All the kinds of tilings we've talked about so far have been periodic. What this means, in a nutshell, is that the pattern of the tiling repeats in two independent directions (a property that also ensures it belongs to a wallpaper group). One way to tell if a tiling is periodic is to construct a lattice – a grid of two sets of evenly spaced parallel lines. The parallelograms into which the lattice is divided are known as period par-allelograms. If the tiling is periodic there'll be a way to lay the lattice over it so that the period parallelograms contain identical blocks known as fundamental domains. By the same token, starting from a fundamental domain we can copy, translate, and paste it, indefinitely, all over the plane to recreate the tiling.

There are infinitely many periodic tilings. There are also infinitely many non-periodic tilings – ones that lack translational symmetry and, therefore, would fail the grid test just described. In the past, mathematicians thought that if you could make a non-periodic tiling from a set of tiles, then you could also make a periodic tiling from the same set. Isosceles triangles, for example, will tile periodically, but can also be arranged in a radial pattern, which, though highly ordered, is clearly non-periodic.

In 1961, Chinese logician and mathematician Hao Wang wondered whether it would always be possible to determine in advance, using a well-defined procedure or algorithm, if a set of tiles can tile the plane. He focused his attention on sets of square tiles whose edges were coloured in various ways. These became known as Wang dominoes. He conjectured that it should be possible to make such a determination, on the assumption that every set of tiles that can tile the plane can do it periodically. A couple of years later, however, one of his students, Robert Berger, showed that this assumption was flawed. Using Wang dominoes, Berger found the first example of an aperiodic tiling – one made from tiles that could be arranged to form a non-periodic tiling but not a periodic one. It was a massively complex affair that involved more than 20,000 tiles. Later, Berger found a set of just 104 Wang dominoes that managed the trick of aperiodicity. Others, including computer scientist and algorithm specialist Donald Knuth, reduced the number still further. Many variations on the Wang dominoes are possible by adding projections and slots but they're all approximately square in form. In 1977 amateur American mathematician Robert Ammann found an aperiodic tiling that used just six square-type tiles. Whether any further reduction is possible using tiles derived from the Wang domino prototype isn't known, although it seems unlikely.

Further progress was made, however, by shifting attention to other types of tile that might force aperiodicity. Leading the charge in this area of research was the English mathematician and mathematical physicist Roger Penrose, best known for his work on the general theory of relativity and cosmology. In the early and mid-1970s, Penrose discovered three different types of aperiodic tiling, which are now named after him. The first, referred to as P1, is made from a pentagon and three other shapes: a 'diamond', a 'star', and a 'boat'. The diamond is a skinny rhombus (a quadrilateral with four equal sides and equal opposite interior angles), the star is a pentagram (having five points), and the boat is a piece of the pentagram (roughly three fifths of it). These shapes have to be put together according to certain rules and are normally shown in different colours.

The other two tilings that Penrose found each use just two different tiles. P2, the best known because it has been talked about widely, is made from a kite and a dart of very specific proportions. These two shapes can be obtained from a single rhombus whose long diagonal is divided in the ratio $1:1/\phi$, where ϕ (phi) is the golden ratio. Alternatively, the kite can be thought of as two joined golden triangles – obtuse isosceles triangles in which the ratio of the length of the equal sides to the length of the third side is one over the golden ratio. The dart, on the other hand, is made from two 'golden gnomons' – triangles whose three angles are in the proportion 1:1:3. The acute angle of the golden gnomon is 36°, which is the same as the apex of the golden triangle.

Without further embellishment, the kite and dart would be able to tile the plane periodically. To avoid this possibility, notches and tabs can be put on the ends of the tiles, or more aesthetically pleasing, coloured circular arcs can be added, with the rule that the tiles must be pieced together in a way that makes matching colours join.

The third type of Penrose tiling, P3, is made of two different rhombi with acute angles of 36° and 72°, respectively. Again, these must be put together in specific ways to avoid periodicity. For instance, they're not allowed to be placed so as to form a parallelogram. A common feature of all the Penrose tilings is local fivefold rotational symmetry. Penrose and, independently, John Conway, proved that whenever the coloured arcs on the tiles closed to form a circle, pentagonal symmetry would be displayed by the entire region around the curve.

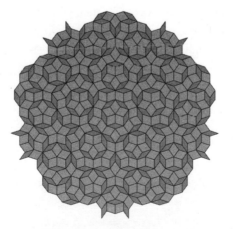

A Penrose tiling (P3) using thick and thin rhombi.

Showing acute business acumen, Penrose filed for a patent on his discovery – and was granted one in 1979 – before announcing it to the world. Some might argue that patenting a natural phenomenon of the universe sets a risky precedent for those engaged in pure research. Others might conclude that, in the eyes of the law, at least, it makes a philosophical point about whether mathematics is discovered or invented. On the other hand, Penrose did invest a great deal of his spare time looking into the problem and it could be said that his efforts, like those of all creative artists, deserve some financial reward.

Sir Roger was not slow in chasing up anyone who breached his patent. In 1997, his wife returned home with some rolls of Kleenex quilted toilet paper, which Penrose quickly spotted were embossed with one of his patterns. The Oxford mathematician reportedly experienced 'shock and dismay' at finding his design displayed in such an unseemly manner. 'He wasn't pleased,' said his lawyer, as reported in *The Wall Street Journal*. Accordingly Penrose and Pentaplex Ltd, the Yorkshire firm that owns the licensing rights to Penrose tilings, sued Kimberly-Clark, manufacturers of the Kleenex brand, for breach of copyright. It demanded, too, that all existing stock of the offensive loo rolls be destroyed and an inquiry made into Kimberly-Clark's profits from the product to assess damages. 'So often we read of very large companies riding rough-shod over small businesses or individuals,' said David Bradley, director of Pentaplex. 'But when it comes to the population of Great Britain being invited by a multinational to wipe their bottoms on what appears to be the work of a Knight of the Realm without his permission, then a last stand must be made.'

Whether embossing bathroom tissue with an aperiodic

pattern rather than a periodic one holds any cosmic significance, we may now never know. Penrose tilings are, however,
immensely fascinating for other reasons. Firstly, there's an
uncountable infinity of them. Secondly, and very surprisingly, all Penrose tilings are alike: every part of any Penrose
tiling is contained infinitely often in every other such tiling.
It's impossible, therefore, to tell from any patch of a tiling
to which tiling overall it belongs. To explain the weirdness
of this the writer and recreational mathematician Martin
Gardner imagined what it would be like if you lived on an
infinite plane tessellated by one of the uncountable infinity
of Penrose tilings:

> You can examine your pattern, piece by piece, in
> ever-expanding areas. No matter how much of it
> you explore you can never determine which tiling
> you are on. It is no help to travel far out and examine
> disconnected regions, because all the regions belong
> to one large finite region that is exactly duplicated
> infinitely many times on all patterns.

The English mathematician John Conway proved a remarkable theorem about matching regions of Penrose patterns.
Suppose the diameter of a certain circular region of one
tiling is d. Starting from a randomly chosen point on another
Penrose tiling, how far away will the nearest identical circular
region be? Conway showed that the distance to the perimeter of the nearest matching region will never be more than
d times half the cube of the golden ratio, or approximately
$2.11d$. The same is true of identical regions on the same tiling:
the distance from perimeter to perimeter is never more than
about double the diameter of the region.

Aperiodic tilings as purely mathematical creations came as a surprise. But this was nothing compared to the shock that scientists got when they found them in the real world. It was pretty much taken for granted that all crystal forms in nature have rotational symmetry of order 2, 3, 4, or 6, and all show extreme regularity in the arrangement of their faces and cleavage planes. But in 1976 Roger Penrose hinted, in a letter to Martin Gardner, that 'quasi-periodic' crystals might be a possibility. Gardner had recently notified Penrose about a new discovery made by Robert Ammann: two rhombohedra that tiled space in an aperiodic way. Penrose pointed out that some viruses take on dodecahedral (12-sided) and icosahedral (20-sided) forms, and that it had always been something of a puzzle how they did this. He added:

> But with Ammann's non-periodic solids as basic units, one would arrive at quasi-periodic 'crystals' involving such seemingly impossible (crystallo-graphically) cleavage directions along dodecahedral or icosahedral planes. Is it possible that the viruses might grow in some such way involving non-periodic basic units – or is the idea too fanciful?

Far from being fanciful it proved to be extraordinarily prophetic. Over the next few years, speculation grew in the research community that crystalline structures based on aperiodic lattices might exist. Then, in 1984, came a sensational announcement. Israeli materials scientist Dan Shechtman and colleagues at the US National Bureau of Standards (where Shechtman was on sabbatical) reported that they'd found an aperiodic structure in electron micrographs of a rapidly cooled aluminium-manganese alloy. The micrographs of what some chemists quickly dubbed 'shechtmanite' showed a clear fivefold symmetry, strongly suggestive of an aperiodic space tiling akin to Penrose tiling. For his discovery of what became known as quasicrystals, Shechtman was awarded the 2011 Nobel Prize in Chemistry.

It took a long time, however, for the reality of quasi-crystals to be widely accepted, so much did it fly in the face of received wisdom. 'It was me against the world,' recalled Shechtman. 'I was a subject of ridicule and lectures about the basics of crystallography.' One of his fiercest critics was two-time Nobel laureate Linus Pauling, who insisted: 'There is no such thing as quasicrystals, only quasi-scientists.'

Today, no one doubts the existence of quasicrystals. Hundreds of types, with different compositions and sym-metries have been identified, based on a variety of metal alloys. The first to be produced were thermodynamically unstable and would revert to an ordinary crystalline form upon being heated. But the first stable ones were discovered in 1987, enabling samples to be produced in large enough quantities for detailed studies that may one day lead to technological applications. After a long hunt for naturally occurring quasicrystals, an international team of scientists finally identified one in the form of a substance given the name icosahedrite. Having the chemical formula $Al_{63}Cu_{24}Fe_{13}$, it was found as tiny grains in a sample collected from an out-crop of the mineral serpentine from the Koryak Mountains in Russia. Analysis showed that it almost certainly came from space aboard a type of meteorite called a carbonaceous chondrite around 4.5 billion years ago, not long after the Earth formed. A geological expedition to the place where it was found identified more specimens of the meteorite, confirming its extraterrestrial origin. The same type of aluminium-copper-iron quasicrystals had previously been made in the lab by Japanese metallurgists in the late 1980s.

There are still many unsolved problems connected with tilings, both in maths and in nature. In the case of Penrose tilings the smallest number of different tiles needed is currently two. Can it be reduced to one? No one has any idea and it remains a fascinating open question.

Another outstanding problem was posed by German geometer Heinrich Heesch in 1968. The so-called Heesch number of a shape is defined as the maximum number of times that a shape can be surrounded by copies of itself (with no gaps or overlaps). Obviously in the case of a triangle, a quadrilateral, a regular hexagon, or any other single shape that can completely tile the plane, the answer is infinity. Heesch's problem was to determine the set of *finite* numbers that can be Heesch numbers, including the largest possible finite Heesch number.

In thinking about this problem, it's helpful to define the Heesch number more precisely. In a tiling, the *first corona* of a tile is the set of all tiles that have a common boundary point with the tile, including the original tile itself. The *second corona* is the set of tiles that share a point with anything in the first corona, and so on. The Heesch number of a shape is the maximum value of k for which all tiles in the k-th corona of any tiling are congruent to that shape. For a long time the record holder for the largest finite value of k was three in the case of a shape found by Robert Ammann, which consisted of a regular hexagon with small projections on two sides and matching indentations on three sides. However, in 2004, Casey Mann, a mathematician at the University of Washington at Bothell, showed that there was an infinitely large family of tiles, consisting of indented and outdented forms of a pentahex (a group of five hexagons) for which the Heesch number is five. This remains the largest finite value known, though it seems likely it will be surpassed in the future.

The Heesch number question seems closely connected to two other famous open tiling problems: does there exist an algorithm for determining whether a shape can tile, and does there exist a shape that can *only* tile aperiodically? Aperiodic tiling seems to act as a barrier to the existence of tiling algorithms, so it isn't expected that both of these problems have the same answer. On the other hand, if no finite Heesch number is larger than some k, then it seems that this could be used as the basis of an algorithm to test whether a shape tiles: simply attempt to fill out a tiling to the $(k + 1)$st corona; if successful, the shape must tile the plane, and if not, the shape doesn't tile.

Although many unsolved problems remain and new ones arise all the time, there have been some remarkable and surprising breakthroughs. Some involve higher dimensions. For example, in 1981, Dutch mathematician Nicolaas de Bruijn proved the extraordinary result that every Penrose tiling by thick and thin rhombi (type P3) can be made by projecting a five-dimensional cubic structure onto a two-dimensional plane cutting through five-dimensional space at an irrational angle.

At the other end of the scale of mathematical sophistication, though no less important, was the discovery made by Marjorie Rice mentioned at the start of this chapter. Despite having no maths training beyond the year she'd taken in high school, she became fascinated by a claim made in Martin Gardner's column in the July 1975 edition of *Scientific American*. Gardner reported that, according to a 1968 proof, the classification of all tessellating convex polygons (those with interior angles of less than 180°) was complete. Rice wondered if the experts had missed something and began sketching shapes on the tile countertops in her kitchen. It wasn't her first private exploration of mathematical puzzles. One of her sons recalled that she'd always been interested in curiosities to do with numbers and geometry, such as the golden ratio and the dimensions of the Great Pyramids. Nor did her lack of mathematical background deter her from tackling the tiling problem. 'I developed my own notation system,' she said, 'and in a few months discovered a new type.' She sent her finding – a new type of pentagonal tiling – to Gardner, who conveyed it to a specialist in the subject for verification. Rice's homegrown technique looked at the different ways that the corners of a pentagon could possibly come together at the vertices of a tiling. Using it she went on to discover four new tessellating convex pentagons and 60 previously unknown types of tessellation based on them.

Rice declined offers to give talks about her discovery and even kept her work secret from her children, though they eventually learned of it as news spread through the academic world and the popular press. She died in July 2017, at the age of 94, having suffered for some years from dementia. Coincidentally, in the same month, French mathematician Michaël Rao published a proof that once and for all completed the classification of convex polygons that tile the plane. It showed that there are 15 types of tessellating pentagon, and no more – including the four uncovered by Marjorie Rice's kitchen doodling. Her achievement was outstanding not just for its ingenuity but because it shows that, even today, it's still possible for someone without training to explore new ground at the frontiers of maths.

CHAPTER 8

Weird Mathematicians

A lot of mathematicians are a little bit strange in one
way or another. It goes with creativity.

– Peter Duren

JAMES WADDELL ALEXANDER II always left his office
window on the third floor of Fine Hall, Princeton University,
open so that he could enter by climbing up the side of the
building. An outstanding topologist who pioneered the
concept known as cohomology as well as the theory of
knots, Alexander was also an expert rock climber. In all
likelihood, he's the only person to have had a strange top-
ological object – Alexander's Horned Sphere – *and* a tricky
ice route in the Colorado Rockies – Alexander's Chimney –
named after him.

Another American mathematician, Ronald Graham, is
best known for his discovery of a ridiculously vast number,
which found its way into the *Guinness Book of Records* for
being the largest number ever used in a mathematical proof.
He's also featured in *Ripley's Believe It or Not* for combining
the talents of a world-class number theorist with those of a
'highly skilled trampolinist and juggler'. Uniquely, Graham

has been a past president of both the American Mathematical Society and the International Jugglers' Association.

Every walk of life has its share of colourful characters and eccentrics but maths seems better endowed than most in this respect. There are top mathematicians, like Alexander and Graham, who stand out simply because they excel in some other, totally different sphere. Then there are those who are so immersed in maths, to the exclusion of almost everything else, that they become detached from the normal world and develop what, to the rest of us, seem oddball traits and personalities. Among the latter was the Hungarian mathematician Paul Erdős, a close friend of Ronald Graham's, who was so prolific that Graham was moved to devise the concept of the Erdős number. If you've ever co-authored an academic paper then you'll probably have an Erdős number, which is the number of jumps needed to connect you with one of Erdős's papers. Your Erdős number is 1 if you're one of the 509 researchers who've actually co-authored a paper with the great man, 2 if you've co-authored with someone who's co-authored with Erdős, and so on.

In a lifetime devoted to maths, and pretty much nothing else, Erdős's output was a staggering 1,525 papers. He had no job or permanent home, and carried around the few possessions he owned in a couple of battered suitcases. Most of his earnings he gave to charity or offered as prizes for solving problems that, for some reason or other, he hadn't yet managed to solve himself. He travelled from university to university, staying with mathematical friends who looked after him and collaborated with him until, after a few days, he'd worn them out with his intense, non-stop intellectual activity. In the last couple of decades of his life, he worked nineteen hours a day, heavily dosed on espresso, caffeine tablets, and amphetamines to keep him permanently alert. In 1979, concerned about his drug use, Graham bet him $500 that he couldn't give up his habit for a month. Erdős promptly did, claimed his reward, and said: 'You've showed me I'm not an addict. But I didn't get any work done … You've set mathematics back a month.' He then went back to popping his Benzedrine pills.

Obsessiveness with mathematics goes back a long way – at least as far as Pythagoras and his followers, two and a half thousand years ago. Not much is known for certain about Pythagoras because none of his writings has survived, and a whole body of myth, some of it amusingly fanciful, has grown up around him. But that he was the head of a secretive school or cult that believed in the migration of souls to new bodies after death, the central importance of numbers, and the 'harmony of the spheres' (music produced by the movements of the Sun, Moon, and planets) is fairly well accepted. It also seems he had a thing about beans, which, along with all forms of meat, he strictly forbade his disciples from eating. According to one account, probably apocryphal, his aversion to trampling on the rights of this humble vegetable led to his undoing. Having been chased from his house by assailants, he came upon a field of beans, which he refused to cross, saying, the story goes, that he'd rather die. At which point, his attackers caught up with him and granted him his wish by slitting his throat.

Equally tragic, and more certain in detail, was the death in 1832 of Évariste Galois, a brilliant young French mathematician. As a teenager, Galois's great ability to solve problems effortlessly in his head, and write down results without filling in the detail of how he arrived at them, exasperated his teachers and held him back academically. That didn't prevent him from doing his own research, however, summarised in a handful of papers, some published posthumously, in which, while searching for solutions to polynomial equations of degree five, he effectively founded the theory of groups.

Things began to go badly wrong for Galois in 1829, beginning with his father's death by suicide. The young man's political activities – he was a staunch Republican, prone to outspoken and hot-headed demonstrations – led to him being thrown into prison on a couple of occasions, though he continued doing maths while behind bars. Shortly after being released from his second spell in gaol he got involved in a duel, perhaps over a woman, though the circumstances remain unclear and it may possibly have been a trap laid for him by political opponents. In any event, Galois was shot in the abdomen during the contest and died the following day, aged just twenty. So sure was he that he'd be killed in the duel that he spent the night before writing down some of his most important mathematical ideas. These notes, together with a handful of unpublished papers, were found fourteen years later by Joseph Liouville, the discoverer of transcendental numbers, who recognised Galois's previously unknown writings as works of genius and brought them to the attention of the world.

Galois's great strength was his ability to leap ahead of others and break into new areas of maths without having to dwell too long on the intermediate steps. But being so far ahead of his time and lax in his proofs frustrated his contemporaries. It was only much later that the importance of his contributions was fully realised. The same could be said, to an even greater extent, of another mathematician, Srinivasa Ramanujan, whose life was also cut short.

Largely self-taught in the early stages of his career, Ramanujan was the most enigmatic and mystical of mathematicians.

He seemed to pluck ideas out of thin air, or, as he would claim, out of dreams as gifts from Namagiri, a Hindu goddess of creativity. Results often came to him, he said, fully realised. After one such episode, he wrote:

> While asleep I had an unusual experience. There was a red screen formed by flowing blood as it were. I was observing it. Suddenly a hand began to write on the screen. I became all attention. That hand wrote a number of results in elliptic integrals. They stuck to my mind. As soon as I woke up, I committed them to writing …

A bust of Srinivasa Ramanujan in the garden of Birla Industrial & Technological Museum, Kolkata.

Working in his spare time while employed as a lowly clerk in Madras (now Chennai) Ramanujan made new and profound discoveries in number theory or, without realising it, rediscovered results that had taken mathematicians in the West centuries to obtain. Even when he arrived at conclusions already known, he'd often do it in a totally original way, by what seemed pure intuition.

He wrote to several leading mathematicians in England in the hope of interesting them in what he'd found but was largely ignored until he sent a letter to G. H. Hardy at Cambridge early in 1913. Hardy recognised some of the Indian's formulae, while others 'seemed scarcely possible to believe'. As far as Ramanujan's theorems on continued fractions were concerned, he 'had never seen anything in the least like them before' but suspected they must be true 'because, if they were not true, no one would have the imagination to invent them'.

Hardy invited Ramanujan to join him and his close colleague, John Littlewood, at Cambridge, but the decision wasn't an easy one for the Indian. He would have to leave behind his family, his 13-year-old wife (by an arranged marriage), and the lifestyle he knew, and lose his Brahmin status, since it was taboo for members of this caste to cross the sea. Ramanujan's mother was at first completely opposed to his going but three months later relented after, she said, being told in a dream by the goddess Namagiri not to stand in her son's way. Accompanied by another Cambridge mathematician, who fortuitously had been staying in India, Ramanujan set sail aboard the SS *Nevasa* bringing with him a suitcase stuffed full of notebooks containing mathematical gems he'd unearthed over the years.

In England, Ramanujan found life difficult – the weather was damp and cold, the culture was alien to him, and he didn't speak the language well. His insistence on following a strict vegetarian diet compliant with Brahmin principles meant he had to cook all his own meals. This he did irregularly and, due to the outbreak of World War I in 1914, without access to some of his customary foods. He effectively became malnourished. On the upside, in Hardy he had a brilliant teacher who set about the delicate task of filling in some of the blanks in Ramanujan's mathematical knowledge while not undermining his confidence and natural freedom of thought. Recalled Hardy:

> The limitations of his knowledge were as startling as its profundity ... It was impossible to ask such a man to submit to systematic instruction, to try to learn mathematics from the beginning once more. On the other hand there were things of which it was impossible that he would remain in ignorance ... so I had to try to teach him, and in a measure I succeeded, though I obviously learnt from him much more than he learnt from me.

For almost three years, despite severe problems in adapting to his new surroundings, Ramanujan prospered academically. In partnership with Hardy, who played a critical role in correcting his proofs and presentation, he published a series of important papers. Filmmaker Matthew Brown, who directed the 2015 film *The Man Who Knew Infinity*, became fascinated by the relationship between Ramanujan and Hardy:

> They are two men so fundamentally different. Ramanujan was a Brahmin Indian from Madras

with no formal education, who believed a formula had no meaning unless it expressed a thought of God. Hardy, on the other hand, was a revered professor at the prestigious Trinity College at Cambridge University and also an avowed atheist. It is an incredible story of how two people were able to overcome their personal differences to form one of the greatest collaborations in the history of mathematics.

Sadly, the partnership ended all too soon. In the spring of 1917 Ramanujan fell seriously ill, possibly with tuberculosis, and was in and out of sanatoria for the rest of his time in England. In 1919 he recovered sufficiently to travel back to India, where it was hoped the more familiar climate and food would help restore his health, but he died the following year, at the height of his mathematical powers, aged just 32.

Even today, researchers continue to sift through the disorganised but fascinating notebooks he left behind, in search of new treasures. Some of his final work, on a subject called mock theta functions, proved, more than eighty years later, to be important in the physics of black holes and string theory. In other cases, scholars are still trying to understand how Ramanujan arrived at his results – or even if they're correct.

The unique powers of this Indian genius raise some interesting questions about the nature of intuition and of mathematics itself. How is it that someone who was so comparatively naive in the subject could make such profound discoveries? What was special about Ramanujan that allowed him to be so much more receptive to mathematical insight than any of his contemporaries? No one would argue that he didn't have a first-class brain: the rapid progress he made under Hardy's tutelage proved that. It wasn't all down to pure insight. But somehow it seems that his religious faith – his belief that formulae and truths about numbers were divine gifts that came to him in visions – opened his mind to the possibility of directly accessing mathematical reality.

Another mathematician whose sudden inspirations often put him ahead of his time was Irishman William Rowan Hamilton, who also did important work in several areas of physics. One of his greatest insights was to treat complex numbers as pairs of real numbers and thereby undercut the prejudice that still existed about imaginary quantities (multiples of the square root of minus one). Extending this approach from the plane to three-dimensional space, as we saw in Chapter 6, he hit upon the notion of special quartets of numbers, which he called quaternions and which are useful for describing rotations in 3D. The idea for them came to Hamilton in a flash one day in 1843, while he was standing on Brougham ('Broom') Bridge over the Royal Canal in Dublin. He recalled:

> The quaternion was born, as a curious offspring of a quaternion of parents, say of geometry, algebra, metaphysics, and poetry ... I have never been able to give a clearer statement of their nature and their aim than I have done in two lines of a sonnet addressed to Sir John Herschel: 'And how the One of Time, of Space the Three, / Might in the Chain of Symbols girdled be'.

Hamilton quite fancied himself as a poet. Through his literary interests, he became friends with both Samuel Taylor Coleridge and William Wordsworth and, like them, wrote stanzas in the Romantic style of the time – though not nearly as well. Wordsworth wanted to be encouraging but, concerned that Hamilton might spend too much time on his poems, gently reminded him that his real talent lay in maths and science.

Hamilton was a classic eccentric and an embodiment of the popular image of an absent-minded professor. Although cheerful, amiable, and courteous in the extreme, he was usually late for appointments and under the illusion that he could explain even difficult topics well to the average listener. The truth is that he wasn't a great success as a lecturer, being prone to wander off topic and ramble on about whatever idea happened to pop into his head. A genius at theorising how to bring order to the mathematical and physical world, he was so absorbed in his studies that he often neglected practical matters. The room where he worked was chaotic and awash with papers, apparently stacked and strewn at random, yet Hamilton could always tell if someone had disturbed the clutter, even slightly. Later in life, his attention to himself and his surroundings went further downhill. Hamilton ignored so many meals while working that, as E. T. Bell put it in *Men of Mathematics*: 'innumerable dinner plates with the remains of desiccated, unviolated chops were found buried in the mountainous piles of papers, and dishes enough to supply a large household.'

At heart, as his poetry suggested, Hamilton was an incurable romantic. Women found him charming, because of his gentlemanly manners, and apparently attractive because of his intellect. But his personal life wasn't always happy. He fell deeply in love with a woman, Catherine Disney, whom he met in 1824 while on a visit to an estate in County Meath. He became infatuated with her and she with him, but he was still a student and Catherine's parents baulked at the

William Hamilton.

idea of her marrying someone with no money and uncertain prospects. Instead they arranged her marriage to a wealthier man, the Reverend William Barlow, fifteen years her senior from a well-off legal family. Hamilton was distraught and briefly contemplated suicide. Poetry became the outlet for his passionate feelings and many of the poems he wrote were about his lost love.

In 1833, Hamilton married Helen Bayly, with whom he had two sons and a daughter. But it was never a happy partnership. Helen was afflicted by various nervous complaints and became a semi-invalid, while Hamilton, still obsessed with Catherine, suffered from bouts of depression and took to drink. In time, Catherine began secretly exchanging letters with Hamilton. Her husband became suspicious,

Catherine confessed to him about the correspondence, and then attempted suicide by taking laudanum. Five years later, she became seriously ill. Hamilton visited her and gave her a copy of his *Lectures on Quaternions*; they kissed at last, and she died two weeks later. Grief stricken, he carried her picture with him ever afterward and talked about her to anyone who would listen. Although he continued his research and even wrote a new book on quaternions, his self-neglect increased and on 2 September 1865 he died after an attack of gout brought on by an orgy of eating and drinking.

As often happens with great thinkers and visionaries, the value of some of Hamilton's work was only fully appreciated by later generations. His quaternions are nowadays used in computer graphics, robotics, and other areas of technology and science that involve rotations in space. Another of his great discoveries has found its way into the theory of the sub-atomic world. Hamilton rewrote Newton's laws of motion in a powerful new form that involves something called the Hamiltonian. This is the sum of the kinetic energies of all the particles associated with a system, plus the potential energy of the particles. The German mathematician Felix Klein saw that the Hamiltonian, together with the so-called Hamilton–Jacobi equation, which relates waves and particles, might be relevant to the new field of quantum mechanics. At Klein's suggestion, the Austrian physicist Erwin Schrödinger looked into the possibility and, sure enough, was able to incorporate Hamilton's work at the heart of his formulation of wave mechanics.

Advanced mathematics is hard, there's no getting away from it. Breaking new ground in maths, especially opening up whole new areas of the subject that no one has thought of before, is even harder – perhaps the most challenging intellectual thing a person can do. Even the best minds can be put under immense pressure by the intensity of the work and the need to focus much of the time on complex, abstract detail. It's often said that there's a fine line between genius and insanity, but it isn't always clear, when great mathematicians crack, whether they do so because of their subject, their psychology, or other circumstances in their life.

It's sometimes claimed that the English mathematician and computer scientist Alan Turing became mentally unstable towards the end of his life because of the stress of his work. But Turing was horribly persecuted, like many others of that time, because of his homosexuality. Despite being a leading pioneer of computing and artificial intelligence, and helping to shorten World War II by breaking secret Nazi codes, he was given a prison sentence in 1952 for 'gross indecency' (practising homosexuality), with the alternative of chemical castration. He accepted the latter. Two years later he was found dead at his home from cyanide poisoning, though whether he committed suicide (the official verdict), or poisoned himself accidentally by inhaling fumes from an experiment he'd been conducting, remains uncertain.

Another researcher whose life ended tragically, possibly because of opposition to his work, though no one knows for sure, was the Austrian theoretical physicist and mathematician Ludwig Boltzmann. A cofounder of statistical mechanics, along with the American Willard Gibbs who developed the subject independently, Boltzmann was prone to extreme mood swings, between elation and depression – what today would probably be classed as bipolar disorder. He also seems to have been extremely sensitive to how others responded to his theories and didn't cope well with criticism. In 1894 he was appointed to the chair of theoretical physics at the University of Vienna, where a year later Ernst Mach took up the chair of history and philosophy of science. Boltzmann and Mach clashed on basic principles of science, Boltzmann arguing that the behaviour of matter could best be explained in terms of numerous collisions of atoms and Mach firmly denying that atoms even existed. Added to this academic rift, the two men disliked each other personally.

In 1900, tired of the discord at Vienna, Boltzmann accepted a position at Leipzig, where he became a colleague of the physical chemist William Ostwald. Unfortunately, although not antagonistic on a personal level, Ostwald proved to be an even more outspoken opponent of Boltzmann's physical theories. It wasn't, at this stage, with the dawning of quantum theory, that Boltzmann was fighting a lone battle or was even in a scientific minority. Far from it: Ostwald and Mach were the only two significant holdouts against the atomic theory of matter. But Boltzmann's sensitive nature left him emotionally vulnerable to attacks on his work, and during one of his bouts of depression he attempted suicide. Although he failed on this occasion, on 5 September 1906, during a summer vacation at the Bay of Duino near Trieste, he hanged himself while his wife and daughter were swimming. He left no note behind in explanation, though his declining health, susceptibility to depression, and philosophical differences with others, or some combination of these, have been put forward as factors in his final action.

The German mathematician Georg Cantor also faced plenty of opposition to his views – more so than Boltzmann – added to which perhaps was the mental pressure of the subject that defined his career: infinity. Cantor studied maths at the University of Berlin under some of the greats of the day, including Karl Weierstrass and Leopold Kronecker. Later his research led him to consider infinity, not just as some abstract concept, but as a new type of number – a 'transfinite' number. What's more, he realised, infinity came in different sizes. He showed that the set of all real numbers is larger than the set of all natural numbers, and, perplexingly, that there are just as many points on a short line as there are on a line, or a plane, or any multidimensional space that extends forever. On reading his proof of this, his compatriot and friend Richard Dedekind said: 'I see it, but I don't believe it.' Dedekind and, later, the Swedish mathematician Gösta Mittag-Leffler, were among the few researchers at the time who offered Cantor support. Several prominent mathematicians vehemently opposed his ideas about infinity, and not just on technical grounds. The leading French theorist Henri Poincaré believed that Cantor's theory of infinite sets would be regarded by future generations as 'a disease from which one has recovered'. Most hurtful of all to Cantor, on a personal level, were the attacks of Kronecker, his distinguished old mentor. Kronecker went out of his way to pour scorn on Cantor's ideas, suppress publication of his work, and block Cantor's ambition of gaining a position at the prestigious University of Berlin. He even went so far as to label him a 'scientific charlatan' and a 'corrupter of youth' for his heretical views. Meanwhile, some theologians were outraged because they regarded his treatment of infinity as a tractable mathematical concept as posing a challenge to the perception of the infinite power of God. He was even denounced as a pantheist. Cantor, a devout Lutheran, totally rejected this accusation. In fact, he maintained that his ideas about infinity had actually come from God.

In 1884, aged 39, Cantor suffered the first of several attacks of manic depression, exacerbated if not induced by the negative reaction of his contemporaries. In between these episodes, he published further mathematical results but increasingly drifted into speculative theorising in other areas. More and more, he spent time engrossed in the philosophical and theological implications of his ventures into the infinite. In another reflection of his unorthodoxy, he argued at length in favour of the Baconian theory – that the plays normally attributed to Shakespeare were actually written by Francis Bacon. He also composed a dialogue between a master and pupil in which the master makes the case that Joseph of Arimathea had fathered Jesus.

Cantor spent the later stages of his life in and out of sanatoria, battling with depression. His final years were pretty miserable, dogged by poverty, ill health, and a general blackness of spirit. However, he did live long enough to see his work vindicated and the likes of David Hilbert and Bertrand Russell praise what he'd done in the highest terms. Regarding Cantor's development of set theory and his explorations of the infinite, Hilbert thought them: 'the finest product of mathematical genius and one of the supreme achievements of purely intellectual human activity'.

The same comment could have been made about the work of another mathematical giant who, in some ways, was the most eccentric of them all. Kurt Gödel was an Austrian-American logician who, in a couple of theorems published in 1931, rocked the mathematical world. He showed that within any system of maths, big and rich enough to be useful in practice, there are bound to be questions that can neither be proved nor disproved. In the theoretical universe in which most mathematicians spend most of their time, for instance, which is based on the so-called Zermelo–Fraenkel axioms, there'll always be problems that can't be solved by any set of rules or procedures.

Gödel was always a bit unusual. As a youngster he was known as 'Mr Why' because of his endless curiosity. His health was never good and seemed to suffer even more following a childhood bout of rheumatic fever, after which he was convinced his heart had been permanently damaged. When Gödel was thirty, a student and Nazi sympathiser murdered logician Moritz Schlick, founder of the Vienna Circle of philosophers to which Gödel belonged. The effect was to unbalance Gödel to the point where he spent several months in a sanatorium and became increasingly paranoid, constantly fretting for instance that he was going to be poisoned.

In 1940, he and his wife, Adele, left Vienna for Princeton, to avoid his conscription into the German army. At the Institute for Advanced Study he formed a friendship with Albert Einstein. So close was their intellectual bond that Einstein, towards the end of his life, remarked that his 'own work no longer meant much and that he came to the Institute merely ... to have the privilege of walking home with Gödel'. Like Cantor before him, Gödel spent more and more time philosophising as he grew older, to the point of putting together a formal argument, in the arcane symbols of modal logic, for the existence of God.

Gödel's morbid fear of being poisoned meant that he ate very little and when he did, only food that had been prepared by his wife. When Adele was hospitalised for six months in late 1977, therefore, Gödel, already skeletal, began to starve. He finally expired from malnutrition on 14 January 1978, his weight a mere 65 pounds.

One of the strangest mathematicians who ever lived wasn't an individual at all but a collective. Nicolas Bourbaki – the name partly based on that of a general, Charles Bourbaki, in the Napoleonic army – was formed in the 1930s by some of the brightest mathematicians in France as a club in Strasbourg. Holding secret meetings, its purpose was to update university lecture courses and texts in the wake of World War I, which had decimated a generation of young talent. The idea started with two maîtres de conferences (the equivalent of lecturers), André Weil and Henri Cartan, at the University of Strasbourg in 1934. Their initial goal was to write a new textbook on analysis to replace a standard work that they'd been using but that was now outdated. Soon about ten mathematicians were meeting regularly on the project. Early on they decided that their work would be communal, without any acknowledgement of individual contributions and the name Nicolas Bourbaki was chosen for the group's nom de plume.

Over the years, Bourbaki's membership varied; some people in the original group dropped out, others were added, and later there was a regular process of addition and retirement (mandatory by the age of fifty). Rules and procedures were adopted that to outsiders often seemed eccentric and even bizarre. For example, during meetings to review and revise drafts for the various books the group developed, anyone could express their opinion as loudly as they wished at any time so that it wasn't uncommon for several distinguished mathematicians to be on their feet at the same time shouting monologues at the top of their voices. Somehow out of this mayhem emerged work of extreme precision, to the point of pedantry and dryness. Bourbaki would have nothing to do with geometry or any attempt at visualisation, and believed that mathematics should distance itself from the sciences. However, despite its tendency to be dull and long-winded, Bourbaki did achieve its goal – to set down in writing what was no longer in doubt in modern mathematics.

The death of the greatest mathematician who never existed was announced in 1968 but not before 'he' had sent in two requests for membership of the American Mathematical Society. The secretary of the Society at the time, John Kline, was not impressed:

> Now, really, these French are going too far. They have already given us a dozen independent proofs that Nicolas Bourbaki is a flesh-and-blood human being. He writes papers, sends telegrams, has birthdays, suffers from colds, sends greetings. And now they want us to take part in their canard.

A strange blend of humour, tragedy, missteps, and dazzling brilliance has accompanied the progress of this strangest of all subjects. Maths *is* weird but its story is made even more fascinating by the long line of colourful characters that have played a part in its unfolding.

CHAPTER 9

In the Realm of the Quantum

I think I can safely say that nobody understands quantum mechanics.

– Richard Feynman

COMMON SENSE AND everyday understanding aren't much use when it comes to grappling with the physics of the ultra small – the world of the quantum. The branch of physics known as quantum mechanics is powerfully counterintuitive. Yet it's described perfectly, and accurately, by mathematics. Interestingly, some of this mathematics was developed much earlier, before anyone knew if there'd be a practical use for it. It's an example of what the Hungarian-American theoretical physicist Eugene Wigner described as 'the unreasonable effectiveness of mathematics in the natural sciences'. But it's also a field that's inspired breakthroughs in maths and might, it's been suggested, eventually form the basis for a whole new domain – quantum mathematics.

Towards the end of the nineteenth century there were few signs that physics was about to be completely revolutionised. On the contrary, most scientists thought that, apart from tidying up a few loose ends here and there, we had all the theory we needed to explain how the universe worked. Newton's laws of motion together with Maxwell's equations of electromagnetism were considered to be the final words in describing how matter and energy behaved. To the late Victorians, who loved machines and technological innovation, all of nature was like a giant clockwork mechanism ticking away in predictable style. There was nothing we couldn't know, they believed, about the details of nature if we took the time to observe it closely enough.

The first cracks in the walls of classical physics came in 1900 when theoreticians were trying to explain the amount of radiation given off by an object as it gets hotter. In fact, we can be quite precise about the moment of crisis: it came around teatime on 7 October. At his home in Berlin, 42-year-old physicist Max Planck, in a moment of inspiration, hatched a formula that precisely matched experimental results to do with something called blackbody radiation.

Max Planck.

A blackbody is an object that absorbs all the radiation that falls on it, whether it be visible light, infrared, ultraviolet, or any other form of electromagnetic radiation, and then reradiates this energy into its surroundings. Nothing in nature is a perfect blackbody, but it's possible to set up apparatus in a lab, consisting of a hot hollow cavity with a small opening, that behaves pretty close to one. Experiments with such apparatus showed that the amount of radiation given off by a blackbody rose gently at low frequencies (long wavelengths), then climbed steeply to a peak, before falling away less precipitately on the high-frequency (short wavelength) side. The peak drifted steadily to higher frequencies as the temperature of a blackbody rose. For example, a warm blackbody might glow 'brightest' in the (invisible) infrared and be almost completely dark in the visible part of the spectrum, whereas a blackbody at several thousand degrees radiated the bulk of its energy at frequencies we can see. Scientists knew this was how perfect blackbodies behaved because their laboratory data, based on apparatuses that were nearly perfect blackbodies, told them so. The sticking point was to find a formula, rooted in known physics, which matched these experimental curves across the whole frequency range.

Matters seemed to be moving in a promising direction when, in 1896, Wilhelm Wien, of the Physikalisch-Technische Reichsanstalt (PTR) in Berlin, devised a formula that agreed well with the experimental data that had been gathered up to that point. The only trouble was that 'Wien's law' had no firm theoretical footing: it had just been tailored to match the observations. Max Planck set about trying to derive it from a basic law of physics – the second law of thermodynamics, which has to do with the entropy, or degree of disorder, of a system. In 1899, Planck thought he'd succeeded. By assuming that blackbody radiation is produced by lots of little oscillators like miniature antennae on the surface of the blackbody, he found a mathematical expression for the entropy of these oscillators from which Wien's law followed.

But then came disaster – disaster, that is, for the great edifice of classical physics. Several of Wien's colleagues at the PTR, Otto Lummer, Ernst Pringsheim, Ferdinand Kurlbaum, and Heinrich Rubens, did a series of careful tests that undermined the formula. By the autumn of 1900, it was clear that Wien's law broke down at lower frequencies – in the far infrared (waves longer than heat waves) and beyond. On that fateful afternoon of 7 October, Rubens and his wife visited the Planck home and, inevitably, the conversation turned to the latest results from the lab. Rubens gave Planck the bad news about Wien's law.

After his guests left, Planck got to thinking about where the problem might lie. He knew how the blackbody formula had to look mathematically at the high-frequency end of the spectrum given that Wien's law seemed to work well in this region. And he knew, from the new experimental results, how a blackbody was supposed to behave in the low-frequency regime. So, he took the step of putting these relationships together in the simplest possible way. It was a guess, no more – a 'lucky intuition', as Planck put it – but it turned out to be absolutely dead on. Between tea and supper, Planck had the formula in his hands that told how the energy of blackbody radiation is related to frequency. He let Rubens know by postcard the same evening and announced his formula to the world at a meeting of the German Physical Society on 19 October.

It was immediately hailed as a major breakthrough. But Planck, methodical by nature and rigorous in his science, wasn't satisfied simply to have the right equation. He knew that his formula rested on little more than an inspired guess. It was vital to him to be able to figure it out, as he'd done with Wien's law – logically, systematically, from scratch. So began, as Planck recalled, 'a few weeks of the most strenuous work of my life'.

To reach his goal, Planck had to be able to add up the number of ways a given amount of energy can be spread across a set of blackbody oscillators; and it was at this juncture that he had his great insight. He brought in the idea of what he called energy elements – little snippets of energy into which the total energy of the blackbody had to be divided in order to make the formulation work. By late 1900, Planck had built his new radiation law from the ground up, having made the extraordinary assumption that energy is transferred not continuously but in tiny, indivisible lumps. In the paper he wrote, presented to the German Physical Society on 14 December, he talked about energy 'as made up of a completely determinate number of finite parts' and introduced a new constant of nature, h, with the fantastically small value of about 6.7×10^{-27} erg second. This constant, now known as Planck's constant, connects the size of a particular energy element to the frequency of the oscillators associated with that element.

Something new and extraordinary had happened in physics, even if nobody immediately caught on to the fact. For the first time, someone had hinted that energy isn't continuous. It can't, as every scientist had blithely assumed up to that point, be traded in arbitrarily small amounts. Energy comes in indivisible bits. Planck had shown that energy, like matter, can't be chopped up indefinitely. It's always transacted in tiny parcels, or *quanta*. And so Planck, who was anything but a maverick or an iconoclast, began the transformation of our view of nature.

You might suppose that such a discovery would have caused an immediate sensation in physics, but no. In 1900 there were some physicists who hadn't even accepted the existence of atoms! For the majority who had there were still many unanswered questions, such as how electrons were distributed inside atoms and the origin of the different spectra of chemical elements. Although Planck's ideas didn't spark an overnight revolution they did attract a growing following to what later became known as the old quantum theory. In this, the fact that only certain values of energy (and some other physical quantities) were allowed was just tacked on to classical physics.

In 1911 came New Zealand physicist Ernest Rutherford's shocking revelation about the structure of the atom. A couple of years earlier, two of Rutherford's colleagues at the University of Manchester, Hans Geiger and Ernest Marsden, had fired alpha particles at a thin gold foil and been astonished to find some of the alpha particles bounced back almost the way they'd come from. Rutherford said it was 'as if you fired a 15-inch shell at a piece of tissue paper and it came back and hit you'. His conclusion: almost all the mass of the atom was concentrated in a tiny nucleus, equivalent in scale to a marble at the centre of a football stadium. The much lighter electrons, he assumed, lay well outside the nucleus. To the amazement of everyone, the atoms of which planets, people, pianos, and everything else are made consisted almost entirely of empty space.

Rutherford painted a picture of the atom as being like a miniature solar system with the nucleus in place of the Sun and the electrons like planets orbiting around it. But there was something obviously wrong with this model. In classical physics, accelerating charges radiate energy. Anything moving along a curved path is accelerating because it's continuously changing direction. If the negatively charged electrons were circling around the nucleus why didn't they quickly radiate away their energy and spiral into the nucleus? If Rutherford's model was correct and, in addition, electrons followed the dictates of classical electromagnetism, all the atoms in the universe ought to collapse in the wink of an eye. Since we're still here, something was missing.

In 1913, the Danish physicist Niels Bohr, who'd also been working in Rutherford's lab in Manchester, brought Planck's ideas on energy quantisation into the picture of the atom. Electrons, he argued, could exist only in certain well-defined energy states within any given type of atom. Providing an electron was in one of these states it wouldn't radiate energy. Only in moving from one energy level to another would it gain or lose energy by a specific amount, by emitting or absorbing a photon – a particle of light. The emission or absorption of photons from transitions between the allowed energy levels of the hydrogen atom, Bohr was able to show, gave rise to the characteristic lines observed in the hydrogen spectrum. Bohr's theory of the hydrogen atom marked the end of the old quantum theory and the beginning of what became known as quantum mechanics.

Progress was slowed by World War I, and the loss of many bright young mathematicians and physicists among the millions slaughtered. But in the aftermath rapid break-throughs were made, especially at Niels Bohr's Institute for Theoretical Physics in Copenhagen and at the University of Göttingen in northern Germany. By the start of 1923, physicists had amassed a huge volume of new data on unexplained

Niels Bohr in conversation with Albert Einstein.

phenomena to do with, for instance, the spectrum of the helium atom and the splitting of spectral lines in a magnetic field. Key players at Göttingen were Max Born, a seasoned professor of physics, and the young Werner Heisenberg, both of whom, in Born's words, 'took part in the attempts to distil the unknown mechanics of the atom out of the experimental results'. Together, Born and Heisenberg worked on a radical new scheme for interpreting physical quantities, such as energies, positions, and speeds. The great epiphany, which brought their ideas together, came to Heisenberg while he was recovering from illness on the North Sea island of Heligoland in the spring of 1925. He later wrote:

> I could no longer doubt the mathematical consist-
> ency and coherence of the kind of quantum mechan-
> ics to which my calculations pointed. At first, I was

deeply alarmed. I had the feeling that, through the
surface of atomic phenomena, I was looking at a
strangely beautiful interior, and felt almost giddy at
the thought that I now had to probe this wealth of
mathematical structures nature had so generously
spread out before me.

By the end of that summer, Heisenberg, Born, and Pascual
Jordan, a contemporary of Heisenberg's at Göttingen, had
developed a complete and consistent theory of quantum
mechanics. Known as matrix mechanics it was intensely mathematical, so much so that few other physicists could properly
understand it. A friend of Heisenberg's, from their undergrad-
uate days, Wolfgang Pauli, criticised it as being 'Göttingen's
deluge of formal learning'. But it was soon to be vindicated.

In parallel with the developments at Copenhagen and
Göttingen, French physicist Louis de Broglie published a
thesis in 1922 suggesting that light could behave either as a
wave or as a stream of particles, but not both at the same time.
He argued that if light, which was normally a wave motion,
could take on a corpuscular (particle) form, then small particles such as electrons could also have wavelike characteristics
associated with them. It fell to Austrian physicist Erwin
Schrödinger to build this concept of 'wave-particle duality'
into a rigorous theory known as wave mechanics. Barely a
quarter of a century after Planck had set physics on a new
course with his quantum hypothesis, there were two, seemingly rival versions of quantum mechanics. For a short while
a fierce debate raged over which was correct. Schrödinger said
he was: 'discouraged, if not repelled' by matrix mechanics.
Meanwhile, in a letter to Pauli, Heisenberg wrote: 'The more
I think of the physical part of the Schrödinger theory, the

more detestable I find it. What Schrödinger writes about visualisation makes scarcely any sense.' But the affair was soon settled. In 1926, Schrödinger himself and, independently, the American physicist Carl Eckart, proved that the two formulations of quantum mechanics, wave and matrix, though wildly different in appearance, were exactly equivalent. Since that time, there have been many other developments in the mathematics of the subatomic world, due to theoreticians such as Paul Dirac (who combined the special theory of relativity with quantum mechanics and predicted the existence of antimatter) and Richard Feynman (famed for his 'sum over histories' interpretation of particle behaviour).

A crucial principle introduced by Heisenberg was the uncertainty principle, which, more than anything, revealed how utterly foreign was this strange realm of the quantum. It indicates the limit to the precision with which certain pairs of quantities connected with a particle, notably position and momentum, and time and energy, can be simultaneously known. The product of the uncertainties in each case must be greater than $h/2\pi$, where h is Planck's constant. The implication is profound: nature puts fundamental restrictions on how much we can know about the state of a particle. For example, the more accurately we choose to measure the position of an electron the greater will be our uncertainty about the electron's momentum. This has nothing to do with the precision of our instruments or techniques. The uncertainty principle arises from an inherent fuzziness of everything around us, including the very stuff of which we're made. On a large scale, things seem clear and distinct. But at the basement level of the world, materiality dissolves and all we're left with is a probabilistic description of events in terms of mathematics.

Nowhere is the interface between physical reality and its mathematical infrastructure more sharply seen than in the realm of the quantum. At the smallest of scales, matter seems to lose its substance; particles, such as electrons, smear out into waves – and not even physical waves but waves of probability. It becomes meaningful to ask: to what extent do things at the quantum level physically exist until and unless they're observed? Do they exist, like pi, only in some abstract Platonic state – a land of possibility – until a measurement or conscious intervention forces them into the open? In the realm of the very small, which we can never directly experience, mathematics is our only guide. The maths moreover is precise, the equations governing the behaviour of matter and energy very specific, whereas the nature of matter and energy itself is uncertain and elusive.

Many mathematicians and scientists, ancient and modern, have commented on the effectiveness of maths in describing the natural world and on the beauty of the equations that underlie physical processes. In the case of the mathematics of quantum mechanics, in particular, the level of precision with which it can predict the probability of something happening is extraordinary. The numbers that come out are among the most precise predictions and verifications in all of science – some to 12 decimal places, which is equivalent to measuring the Earth–Moon distance to within a hair's width. It's almost as if there's a reversal of roles of maths and materiality as we descend to smaller and smaller scales. In the everyday world we inhabit, we see and feel concrete objects. We're aware of tangible things whose physical state and condition we can measure to whatever degree of accuracy we please. To be sure, maths is there in the background, ever-present, but it forms an invisible infrastructure that guides the movement of planets, the flight of birds, and the fall of stones. As the graininess of matter at the atomic and subatomic level comes into view, however, maths and matter seem to swap places. Particles dissolve into mere waves of probability and the only certainty becomes the equations that orchestrate, in exquisite detail, the bizarre goings-on in a domain where common sense fails and our very understanding of what it means to be real is thrown into doubt.

So utterly different and unexpected are the rules of the quantum domain that mathematicians see in them an exciting opportunity. Quantum mechanics provides a rich context for the development of new mathematics. Could its strange and unique logical structure, once fully grasped, form the basis for a whole new branch of the subject? Dutch mathematical physicist Robbert Dijkgraaf, who is director of the Institute for Advanced Study in Princeton, believes that 'quantum mathematics' may be the ultimate product of one of the most curious aspects of quantum mechanics. Unlike in classical physics, where moving objects follow definite paths, in quantum mechanics, in going from one point to another, it's as if a particle explores all possible paths at once. The maths of this weird spreading-out assigns a probability to the chances of the particle going down any one of these paths and then sums all the options to give a probability distribution. The likeliest path (though not necessarily the one taken) is the solution that classical, Newtonian physics would give. The 'sum-over-histories' approach, points out Dijkgraaf, has much in common with a branch of modern maths known as category theory. Categories, in maths, are collections of objects related by common algebraic properties. They include sets, rings, and less well-known ensembles of elements, all of whose relationships can be indicated by arrows. What category theory and the sum-over-histories model of quantum mechanics have in common is their holistic, bird's-eye view, where what matters is not the individual (element or particle) but the totality of what's possible.

A striking example of how physics can inform maths, rather than the other way around, concerns an esoteric topic in geometry known as Calabi–Yau spaces. Don't bother trying to conjure them up in your imagination: they exist in six dimensions and emerge as solutions to the equations of Einstein's general theory of relativity (our best current theory of gravity). They're also central to string theory, which attempts to get to grips with some crucial unanswered questions in particle physics. String theory only works if we allow that the space around us has more than the three dimensions with which we're familiar. Calabi–Yau spaces offer a convenient way of curling up these extra dimensions, required by string theory, so small that they're far beyond our ability to detect.

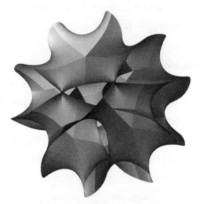

In string theory this shape is known as the Calabi–Yau quintic. It's a candidate for the wrapped-up hidden six dimensions of 10-dimensional string theory.

Mathematicians can classify Calabi–Yau spaces, simply speaking, in terms of how many curves can be wrapped around them, like wrapping an elastic band around a cylinder. But establishing this number of curves is extremely hard. In the case of the simplest Calabi–Yau space, known as the quintic, the number of degree-one curves (lines) that can be fitted on the space was found to be 2,875. German mathematician Hermann Schubert made this discovery in the 1870s, but it wasn't for another century or so that the corresponding number of degree-two curves was established – 609,250. Then a group of string theorists challenged some of their mathematical colleagues to figure out what the number of degree-three curves would be. Meanwhile, the physicists came up with their own solution, using a sum-over-histories technique rather than pure geometry, to calculate not just the number of curves of degree-three but of *any* degree. As Robbert Dijkgraaf has pointed out: 'A string can be thought to probe all possible curves of every possible degree at the same time and is thus a super-efficient "quantum calculator".' The team of string theorists, led by English physicist and mathematician Philip Candelas, came up with a figure of 317,206,375 for the number of degree-three curves. This was completely different from the answer given by the geometers, who obtained theirs after running a complicated computer program. So confident were the string theorists in the general formula they'd devised that they suspected an error had crept into the mathematicians' program. Sure enough, when the geometers checked, they found they'd made a mistake.

Physicists telling mathematicians that they've got their sums wrong is virtually unheard of in the scientific world. Almost always it's the other way around: physics is usually informed by maths. This startling turnaround reveals the potential of this new physics of string theory – a particular branch of quantum mechanics – to cast light not just on previously unknown science but also on unexplored mathematics.

In another surprising development, the most fundamental formula in quantum mechanics, Schrödinger's equation, has turned out to be extremely useful in game theory. This is the branch of mathematics that deals with how players choose a strategy to achieve a goal, such as improving their survival chances, achieving higher profits, or winning at an actual game. When a large number of players is involved, researchers often model the scenario with what's called a mean-field approach. This effectively considers all the players involved as a group and averages over their combined behaviour in arriving at an optimum. Recently, game theorists have found that they can carry out the mean-field method in much the same way as quantum physicists have been using Schrödinger's equation for the best part of a century.

Igor Swiecicki from the University Paris-Saclay and colleagues were studying a certain class of mean-field games using how fish behave in schools as an example. In a crowd of hundreds or thousands of fish it's impossible to take account of how each individual moves. One way to tackle the problem is to use numerical simulations based on the average density of fish in different parts of the school. But this method doesn't take account of the underlying mechanisms that drive the behaviour. A more insightful approach is to assume that each fish will swim in a manner that minimises something called the cost function. This function takes into account, for instance, the energy that the fish has to use and the survival edge the fish gains from swimming in a group to confuse predators. The derived mean-field game equations, it turns out, look a lot like the Schrödinger equation and when solved, give a result that matches the results from numerical simulations.

Quantum physics may shed light on many other areas of maths in the future. But it's also limited by the same principles that govern the whole of mathematics. In the 1930s, Austrian-born logician Kurt Gödel rocked the mathematical world with his discovery of the incompleteness theorems. These showed that, within any system of maths, there'll always be true statements that can't be proved. It took until 2015, however, for physicists to find an example of the incompleteness theorems at work in science.

An international team of researchers was studying if and at what point, as it's cooled, a semiconductor material becomes superconducting. The key factor is the spectral gap – the energy needed to transfer an electron in the substance from a low energy to a higher energy state. If this gap closes up, the material can suddenly switch to a completely different state and become superconducting. But when the researchers applied sophisticated maths to the problem, which included a complete description of the material in terms of quantum mechanics, they were shocked by what they found. Determining whether a spectral gap exists or not turned out to be undecidable. It's a result that has serious implications because it shows that even a full knowledge of a material's microscopic properties isn't enough to predict how it will behave on a larger scale.

The discovery may even limit progress in particle physics. One of the most important unsolved problems in both maths and physics is the Yang–Mills mass gap conjecture. This concerns whether the Standard Model, the scheme that describes the fundamental particles of matter, itself has a spectral gap. Experiments carried out using giant accelerators and lengthy supercomputer calculations suggest that it does. The Clay Mathematics Institute includes the mass gap conjecture as one of its seven $1 million Millennium Prizes for the first confirmed proof based on the equations of the Standard Model. Whether the undecidability of spectral gaps in general stands in the way of specific cases being solved remains to be seen. But there's some good news. One of the reasons the undecidability arises is the weird behaviour of the models used to represent materials at the quantum level. But this behaviour, though impossible to analyse, suggests some bizarre and fascinating physics may be waiting to be found. The addition of a single particle, for instance, might, in some circumstances, alter the properties of a whole lump of matter, with potentially explosive implications for technology.

CHAPTER 10

Bubbles, Double Bubbles, and Bubble Troubles

I wonder how much it would take to buy a soap bubble,
if there were only one in the world.

– Mark Twain

EVERY YEAR SINCE 1825 (except for the war years 1939–42), the Royal Institution in London has held a series of Christmas Lectures for children on a topic of scientific interest. In 1890, the lectures were delivered by physicist Charles Boys on the subject of soap bubbles. 'I hope', he said in his introduction, 'that none of you are yet tired with bubbles, because, as I hope we shall see during this week, there is more in a common bubble than those who have played with them generally imagine.'

Bubbles are child's play but we never grow tired of them. The way they drift unpredictably, rise on currents of air, and slowly fall before bursting makes them fun to watch. Their shifting iridescent colours add to their beauty. And the forms they make when they stick together are fascinating. From an early age we know what they look like when single or grouped in clusters. David's four-year-old grandson has no trouble recognising the shape formed when two bubbles meet. It seems trivial. Yet this problem – the shape formed by two conjoined bubbles – proved baffling to mathematicians for a long time, despite the fact that we've all seen it many times. It led to what's known as the double bubble conjecture, which was finally proved as recently as 2002. Other problems to do with bubbles and soap films remain unsolved to this day.

Conventional bubbles are nothing more than a soap film wrapped around a pocket of air. Two layers of soap molecules form the inside and outside surfaces of the film and are separated by a thin layer of water. A bubble – until it bursts – is airtight: no air can enter or leave it. Even if it isn't popped on purpose or by touching something that breaks the film, it will burst spontaneously when the water between the layers of soap molecules evaporates. Blown on a cold winter's day, bubbles tend to last longer because the evaporation is slower and the bubbles may even freeze.

The key to understanding the shape of a bubble is surface tension – the force that acts on its surface like an elastic skin. Surface tension comes about because of the cohesive (attractive) force between molecules of a liquid. Inside a body of liquid a molecule is pulled equally on all sides by its neighbours, so there's no overall force. But at the surface, molecules are pulled only sideways and downwards, which has the effect of making the surface appear as if it has a skin. In reality, it's more accurate to say that surface tension makes it harder to move an object through the surface than to move it when completely immersed but, for most purposes, it works well enough to imagine that there's an actual skin.

It's a common misunderstanding to suppose that the reason you can't blow bubbles with water alone is that the surface tension of water isn't great enough and that soap increases it. In fact, the opposite is true. Adding soap *decreases* the surface tension. Water bubbles burst almost the instant they form, for a couple of reasons: the surface tension is too great, causing them to tear themselves apart, and evaporation from the surface of the bubble makes them too thin so that they pop. Soap molecules help bubble formation because each consists of a long hydrophobic ('water-fearing') tail made of carbon and hydrogen atoms and a hydrophilic ('water-loving') head made of oxygen and sodium. In a soap-and-water solution, the hydrophobic tails of soap molecules move as far away as possible from the water and so end up at either the inner or the outer surface of a bubble. Meanwhile, the hydrophilic heads project into the water sandwiched between the two layers of soap molecules, which increases the separation of the water molecules and thereby reduces the attractive force between them. Result: the surface tension is lowered. What's more, because the water is partly protected by the soap films it evaporates more slowly.

If kept airborne, bubbles typically last 10 or 20 seconds before they burst. But their lifetime can be greatly extended by keeping them in a sealed container in which the air is saturated with water vapour to reduce drastically the rate of evaporation. Eiffel Plasterer, from Huntington, Indiana, became fascinated with bubbles as a physics teacher in the 1920s and went on to become well known for his bubble-making shows and demonstrations. His prime-time TV appearances included one on *Late Night with David Letterman* in which he managed to form a complete soap bubble around the host. He also holds the world record for soap bubble longevity, having once blown a bubble in a sealed jar that lasted for just 24 days short of a full year!

Bubbology attracts more than its fair share of enthusiasts and master exponents who constantly vie to outdo each other. Matěj Kodeš of the Czech Republic currently tops the list for enclosing the largest number of people within a single bubble – an astonishing 275. He also succeeded, in 2010, in forming a bubble around a six-metre-long truck. Canadian citizen Fan Yang is distinguished for having blown the largest number of bubbles – 12 – one inside another, like a Russian doll. Sam Sam the Bubble Man (aka Sam Heath) of the UK boasts three records: the most bounces by a bubble (38), the longest chain made from interlocking bubbles (26), and the largest frozen soap bubble, with a volume of 4,315 cubic centimetres. As for making the largest-ever free-floating soap bubble, that honour goes to American Gary Pearlman who, in 2015, created a monster with a volume of about 96.2 cubic metres.

Of course, giant bubbles wobble all over the place like an underset jelly on the tray of a nervous waiter. Small ones,

A giant soap bubble.

however, maintain a constant shape, which, as everyone knows, is a sphere. No other shape encloses a given volume with a smaller surface area. A sphere with a volume of 10 cubic centimetres has a surface area of 48.4 square centimetres. The five Platonic solids, tetrahedron (4 sides), cube (6 sides), octahedron (8 sides), dodecahedron (12 sides), and icosahedron (20 sides), with this same volume of 10 cubic centimetres, have surface areas of 71.1, 60.0, 57.2, 53.2, and 51.5 square centimetres, respectively – the area reducing as the shape more and more closely approximates a sphere. Like all things in nature, bubbles tend towards the lowest energy configuration possible. In so doing they minimise the forces of tension in the soap film, which, in turn, involves minimising the surface area that encloses a given volume. The logic, and physics, behind why bubbles are spheres isn't hard to follow. But proving, mathematically, that the sphere is the surface with the minimum area for a given volume is surprisingly hard. In fact a complete proof came as recently as 1884.

It's easier to start with the equivalent problem in two dimensions: what's the curve of the shortest perimeter that encloses a given area? In legend, it's a question to which Queen Dido must have given plenty of thought before she asked the Berber king, Iarbas, for as much land as she could encompass using a single ox hide. Iarbas was happy to grant her wish, comfortable in the knowledge that he wouldn't miss the tiny patch of land that a mere animal skin could cover. The ingenious Dido, though, took the ox hide, cut it into incredibly thin strips, and arranged the strips into an enormous circle big enough to enclose the future city of Carthage. It's hard to imagine that Dido could have done better than choose a circle to maximise her territory. But it's one thing to have a gut feeling that a circle encloses the greatest area for a given perimeter – or, equivalently, that it's the shape with the smallest perimeter for a given area. It's quite another to actually prove it.

Progress towards such a proof was made by Swiss geometer Jakob Steiner. He found various properties that the maximal shape must inevitably have. For instance, it must be convex because if it were concave it'd be possible to reflect the concave part in a straight line, producing a shape with the same perimeter and an even larger area. Using a number of such arguments he concluded that the maximal shape had to be a circle. But his conclusion had one flaw. While Steiner showed that if any shape with the smallest perimeter existed it must be a circle, he didn't show that any such shape necessarily existed in the first place! To understand how such a situation might arise, suppose you say, for instance, that you can 'prove' that the largest positive integer is 1. Start with any supposed largest positive integer n. Then, if n isn't equal to 1, $n^2 > n$, so n wasn't the largest positive integer to begin with. Therefore, n must equal 1. The flaw in the argument, of course, is to assume that there's a largest positive integer to begin with, when there isn't.

In the case of the curve of maximal area, Steiner was correct: there does exist such a curve and, as he proved, it's the circle. But a proof that such a shape exists had to wait for other mathematicians. Many proofs, using many different methods, were put forward. In 1884, a proof was found that the sphere is the surface of minimal surface area that encloses a given volume, and in 1896 German mathematicians Hermann Brunn and Hermann Minkowski generalised this to all higher-dimensional spheres. Nevertheless, these remained special cases. We still didn't know what happened in more complicated cases with more conditions.

In the nineteenth century, Belgian physicist Joseph Plateau came up with a number of laws that could be applied to the shape of bubbles. His first law is that soap films are made of smooth surfaces and his second that the mean curvature is constant throughout each individual film. His third law states that when soap films meet, they always do so in threes at an angle of 120°. This rule applies to when two bubbles are joined because, in this case, there's a meeting of three soap films: one for each bubble, and one for the boundary separating the bubbles from each other. If one of the bubbles is larger, the boundary will curve inwards towards the larger bubble in order to satisfy the third law. Plateau's fourth law is that where three faces meet at 120°, the edges of these faces will themselves meet in fours with angles of approximately 109.5° – the tetrahedral angle. It's so called because if you draw lines from the vertices of a tetrahedron to the centre this is the angle at which they'll meet. Any other pattern of bubbles, Plateau found, will be unstable and will quickly reorganise itself so as to satisfy the laws.

Plateau also considered what would happen if various boundary conditions were imposed. For instance, if a bubble lands on a table it'll form a hemisphere. The angle between the bubble and the table doesn't have to be 120° in this case because the table isn't a soap film and so the total area of the enclosed surface doesn't need to be minimised. If you had a tetrahedral wire frame and dipped it in soap, there'd be six soap films, one for each edge, all pointing inwards, meeting each other at the tetrahedral angle in the four lines that connect a vertex to the centre.

Plateau didn't derive his laws mathematically but instead from a lengthy series of observations – all the more impressive because he'd begun to go blind at the time. What led to his loss of sight isn't certain but it may have been connected with his tendency, as a young man, to perform risky optical experiments. He's known, for instance, to have once stared directly at the Sun for 25 seconds, to see what impressions it would leave on his retina.

Although Plateau was confident enough to state his laws, based on experimental evidence, he didn't know how to prove them. To do so was much tougher than the problem of establishing the minimal surface bounding a given volume, because multiple bubbles were involved each with their own volume. In fact it was only in 1976 that the American mathematician Jean Taylor finally proved that the rules proposed by Joseph Plateau always hold for minimal surfaces. Any surface, she showed, that has a minimum area while satisfying constraints on volumes will conform to Plateau's laws.

Taylor's proof was an important step along the way to proving one of the greatest open problems to do with minimal surfaces: the double bubble conjecture. According to this, the shape that encloses and separates two distinct volumes *and* has the least possible surface area is a standard double bubble – three pieces of spheres meeting along a circle at 120° angles. But her proof didn't resolve the problem entirely. Other configurations were still possible, and all of them had to be ruled out. One such configuration consisted of one bubble being peanut-shaped and the other forming a doughnut-shaped ring around its middle. If it satisfied Plateau's laws – which it did – it remained a possibility. In 2002, however, four mathematicians (Michael Hutchings, Frank Morgan, Manuel Ritoré, and Antonio Ros) finally laid the double bubble conjecture to rest, proving that our intuitions as children were correct and the shape formed by two bubbles is mathematically optimal.

In optimisation problems, points in two-dimensional space where lines meet at 120°, and the 3D analogue where four lines meet at 109.5°, are very common. For instance, take the following easily stated problem:

A, B, and C are three corners of a triangle. Find the point P so that the distance PA + PB + PC is minimised.

First, note that if a point is moved around inside a triangle, the distances to the various corners will increase or decrease. Overall, though, the differences won't cancel out and there'll always be one unique point where the sum of the distances is as small as possible.

If ABC is equilateral, then the obvious – and correct – answer to our problem is the centre of the triangle. For a triangle where all the sides are of different length, though, the solution isn't so clear. For one thing, there are different ways of finding the centre of a triangle. You could take the medians (lines from one corner to the midpoint of the opposite side) and find that they meet at a single point, known as the centroid. This is also, by the way, the centre of mass of the triangle: if you made the triangle out of, say, wood, of uniform thickness and density, and placed a pin under the centroid, the triangle would balance on the pin. You could also take the point that's an equal distance from A, B, and C: the so-called circumcentre. (It's also the centre of the circle passing through A, B, and C.) The angle bisectors (lines which pass through a corner and divide the angle into equal parts) meet at the incentre, and the altitudes (lines which pass through a corner and are perpendicular to the opposite side) meet at the orthocentre. These four are by far the most common triangle centres, and you might expect that the point P must surely correspond to one of them. But, in fact, it doesn't. The point P turns out to be something called the Fermat point, which is rarely heard about. The Fermat point is the point P at which the angles between PA, PB, and PC are all 120°. One way of constructing it is to draw an equilateral triangle on each of the sides, and then, from the third corner of each equilateral triangle, draw a line from it to the opposite corner of the original triangle. These all meet at the Fermat point. There's one oddball situation, which happens if any of the angles in the original triangle are greater than 120°. In this event the construction gives rise to a point that's outside the triangle itself, so obviously it can't be the optimal point. In this case, it turns out, the Fermat point is the corner at which the angle is greater than 120°.

Reassuringly, it's possible to confirm by experiment this mathematical conclusion: that the point P which minimises the distance PA + PB + PC is also generated by the corresponding physical process. Plateau constructed wire frames and dipped them in soapy water. The 2D equivalent in this case would be two glass plates held closely together but not touching (to represent two-dimensional space) and three small metal rods that are fixed to the glass plates (to represent A, B, and C). If you dip this in soapy water and then remove it, there'll be three 'lines' of soap films, and you'll find that they meet at the Fermat point.

Plateau also tested what occurs when a cubical wire frame is dipped in soapy water. In this case, the soap films appear to form a smaller cube inside the main cube, but joined to the main cube by other soap films. A perfect cube, however, won't satisfy Plateau's laws, so the surfaces are slightly curved. In particular, the small cube bulges outwards slightly. The size of the small cube can vary depending on how much air was trapped by the soap films when they formed.

It's fascinating that all that's needed to explore minimal area surfaces – even complex ones that would be hard to calculate in advance – are wire frames of your choosing and a good bubble mixture. (Any number of recipes are available, mostly involving water, washing-up liquid, and some strengthener such as glycerine.) Two circular hoops, for instance, dipped in soap solution and removed give rise to an interesting surface between them. Here no volume needs to be enclosed; instead, all that's required is that the soap film has both hoops as boundaries. You might guess, before doing the experiment, that a cylindrical shape would minimise the surface area of film between the hoops best. But the actual shape obtained curves inwards, becoming narrowest in the middle, before curving outwards again. It's known as a catenoid and is the surface produced by rotating a catenary – the curve produced by supporting a light string at both ends.

Plateau's laws can be applied not just to two bubbles, but to entire foams of bubbles. Although a foam may look, at first glance, to be extremely random, it is in fact highly constrained. Every bubble in it must obey Plateau's laws and, if no bubbles are popped, all bubbles have a specific volume of air that they must contain. A foam is formed by the configuration that, once again, minimises the total surface area while enclosing the specified volumes. Foams appear naturally in places other than the typical soapy water. For instance, in the human body, bones consist primarily of a hard outer layer (compact bone), a softer inner layer (spongy bone), and bone marrow. Spongy bone has a foam-like structure, although in this case the structure is porous. (Instead of there being enclosed bubbles, the bubbles are open and the structure consists of a network corresponding to the edges of the foam.) It's thought this foam-like structure ensures that the bone is flexible rather than brittle.

Foam served as the inspiration for the Beijing National Aquatics Centre, used in the 2008 Olympics. This building, nicknamed the Water Cube (though it's cuboidal rather than cubic in shape), has a pattern that resembles a section of foam. It initially appears realistic, in that it has the irregularity we expect from a foam, but the trained eye can spot discrepancies. For instance, there are bubbles that are rectangular or triangular in shape, which break Plateau's law that all angles between bubbles must be 120°. These bubbles appear out of place in comparison to the rest of the structure. Perhaps the architects weren't aware of Plateau's laws or, if they were, chose to ignore them at certain points for aesthetic or practical reasons.

Foams have two-dimensional equivalents as well. What is the best way to divide a plane into regions of equal area, if you want to minimise the total perimeter? (Of course, the total perimeter will inevitably be infinite – to fix that problem, take the total perimeter over a large region of the plane and divide by that region's area to get the 'average' perimeter.) The answer, again, is intuitively obvious but hard to prove – a honeycomb pattern. The honeycomb pattern is the most efficient way to have cells with the same area that minimises the beeswax required, which may explain why bees use it. Everyone intuitively knew it was true, and we could trivially prove that it's the best regular tiling (there are only three regular tilings: the triangular tiling, the square tiling, and the hexagonal tiling) but we couldn't prove it in general, for irregular tilings. It isn't known when the conjecture was stated. Our earliest source that mentions it is from 36 BCE, from Marcus Terentius Varro, but it's believed to have been considered even before then. By contrast, its proof came in 1999, by Thomas Hales, making it one of the longest-standing open problems in mathematics. The main difficulty came in having to consider every irregular tiling as well as just the regular tilings, and even consider tilings with curved sides, or many different types of tiles, each with the same area.

The three-dimensional version of the Honeycomb Conjecture is even harder. In fact, it's so hard that it's still an open problem. In 1887, Lord Kelvin, after whom the Kelvin temperature scale is named, asked the following: what's the most efficient way to divide three-dimensional space into regions of equal volume and minimal surface area? Kelvin thought he knew the answer but couldn't prove it. He started with the truncated octahedron, a shape formed by slicing the corners off an octahedron, producing a shape with eight hexagonal faces and six square faces. This shape can tile three-dimensional space. Kelvin realised that it can be improved upon if the faces are slightly curved to conform to Plateau's laws but he conjectured that this – the curved truncated octahedral tiling – was the most efficient tiling of all. For over a century no one managed to improve upon Kelvin's tiling. Then, in 1993, Trinity College Dublin physicist Denis Weaire and his student Robert Phelan finally succeeded. The Weaire–Phelan structure is much harder to describe and consists of two types of tiles, a truncated hexagonal trapezohedron and a pyritohedron, and once again the faces are curved so that they follow Plateau's laws. The Weaire–Phelan structure is only a slight improvement on Kelvin's structure – it has 0.3 percent less surface area – but in the form of foams it does appear naturally. In fact, it was discovered while investigating computer simulations of foams. What we still don't know is if the Weaire–Phelan structure is optimal or whether, someday, an even more efficient tiling will be found.

In nature, bubbles and foams form some of the largest and smallest structures that scientists have ever made or contemplated. On a miniature scale are nanofoams, a type of porous material in which a majority of the pores are less than 100 nanometres (billionths of a metre) across. One of the best-known examples is aerogel, sometimes referred to as frozen smoke, because it's incredibly light and almost mist-like in appearance. Nanofoams of various compositions – carbon, metal, and glass among them – have unusual physical properties that may find future applications as incredibly slender wires, efficient catalysts, and energy storage devices.

At the other end of the size range are bubbles on an astronomical scale. These may be blown by powerful winds from the surface of hot, young stars as they smash into the surrounding interstellar medium. Larger cosmic bubbles, known as superbubbles, hundreds of light-years across, may be formed as the result of multiple stellar explosions or supernovae. In fact, our own solar system lies near the centre of such a superbubble, which appears to be the result of several supernovae that exploded in the past some 10 to 20 million years ago.

CHAPTER 11

Just for the Fun of It

A lack of seriousness has led to all sorts of wonderful insights.

– Kurt Vonnegut

IF CHILDREN LEARN best when they're playing and having fun, then recreational maths should be taught in every school. To those who've endured endless hours learning times tables, solving equations, and finding missing angles, 'recreational maths' may sound like an oxymoron. But anyone who enjoys such pastimes as Sudoku, logic puzzles, or Rubik's Cube is having fun doing maths, often without realising it. What's more, playful maths – even puzzles and games with simple rules – can lead to important breakthroughs in mathematics as a whole.

Maths just for the fun of it has been around almost as long as maths itself. Certainly, puzzles to do with numbers, shapes, and logic have been devised for entertainment and as mental challenges since the time of the ancient Greeks. One of the oldest known classic puzzles, the loculus of Archimedes, is to recreate a square from 14 different triangular or quadrilateral pieces. There are many different ways to do this and it was only in 2003 that all the solutions were finally enumerated using a computer program. American mathematician Bill Cutler showed that, barring rotations, reflections, and permutations of identical pieces, there are 536 distinct solutions. The loculus is an example of what's known as a dissection problem. Puzzles of this kind, like jigsaws, don't call for any knowledge of maths (although it may help!) and so can be tackled by anyone.

Another old dissection problem is the tangram, consisting of seven pieces – five triangles of various sizes, one square, and one parallelogram – which come from cutting up a square. The goal is to make a specific shape, of which thousands are possible, given only the outline or silhouette. All the pieces must be used without any overlaps. The tangram seems to have originated in China hundreds of years ago. It was eventually brought to Europe by trading ships in the early 1800s and soon caught on as a popular pastime. Napoleon, Edgar Allan Poe, and Lewis Carroll were all fans of the game and Carroll, in particular, helped revive interest in it in England in the late nineteenth century when he used the pieces to create illustrations of the characters in the Alice books.

Of the same genre is a puzzle, dating to the start of the twentieth century, which is surprisingly difficult considering that it involves just four pieces. The pieces of the T-puzzle may be rotated or flipped over, but not overlapped, in order to make a symmetric capital T. In fact, two different symmetric capital T's can be made from the pieces as well as two other symmetric shapes, including an isosceles trapezoid.

A solid dissection puzzle, similar to the tangram but in three dimensions, was invented by Danish mathematician, inventor, and poet Piet Hein in 1936 following a lecture on quantum mechanics by Werner Heisenberg. The German physicist had been describing a space sliced into cubes when, in a moment of genius, Hein grasped that the result of combining all seven of the irregular shapes that can be made from no more than four unit cubes joined at their faces is a single larger (3 × 3 × 3) cube.

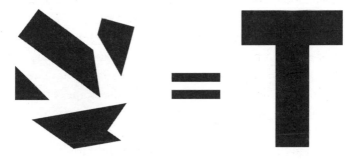

The T-puzzle.

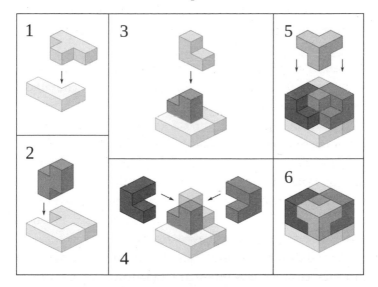

The Soma Cube.

The pieces of the Soma Cube consist of all possible combinations of three or four unit cubes, joined at their faces, in such a way that at least one inside corner is formed. Said Hein:

> It is a beautiful humour of nature that the seven simplest irregular combinations of cubes, can be recombined to the cube again. The multitudes of unity is again producing unity. This is the world's smallest philosophic system, and that surely must be an advantage.

As in the case of many other novel and entertaining ideas in maths, Hein's discovery was put in the world spotlight by Martin Gardner's 'Mathematical Games' column in *Scientific American*. Three years later, in 1961, all 240 possible ways of assembling the 3 × 3 × 3 Soma Cube from its component pieces were determined by British mathematicians and, at the time, fellow Cambridge students, John Conway and Michael Guy. Conway went on to invent a larger, 5 × 5 × 5 form of the Cube, with 18 pieces, known simply as the Conway puzzle, while Hein was rewarded by the commercial success of his game, first as an elegant rosewood edition by a Danish company, marketed in the States by Parker Brothers, and later as a cheaper plastic version.

Hein, Conway, and Gardner are among recent champions of recreational maths who have shown how the fun side of mathematics connects seamlessly with its more serious, academic aspects. Conway is an eminent mathematician who's made important contributions to number theory, knot theory, geometry in various dimensions, and group theory. Hein investigated the superellipse and was famed as an inventor and poet as well as a creative deviser and solver of puzzles.

Gardner was the greatest populariser of maths in modern times, highly respected both among the maths community and the public at large, for bringing fascinating and often new discoveries in the subject to widespread attention.

But, as we've seen, the playful side of maths has a long history. Another brainteaser attributed, like the loculus, to Archimedes is the cattle problem. Again, this wasn't solved in full until quite recently. The answer to it was found in 1880 but turned out to be so huge that it wasn't calculated and printed out accurately until 1965 – once more, as in the case of the loculus, with the help of computers. The difference between the loculus and the cattle problem, though, is the level of difficulty. Anyone can have a go at shifting pieces of a puzzle around to fit them together in a particular way. But the cattle problem is enough to put off most people at a glance. More than two thousand years ago, Archimedes posed it as a challenge to the clever mathematicians of Alexandria, headed by Eratosthenes. He starts off: 'Compute, o friend, the number of oxen of the Sun, giving thy mind thereto, if thou has a share of wisdom.'

Then, paraphrasing a bit, the problem runs along these lines: The sun god has a herd of cattle consisting of bulls and cows, one part of which is white, a second black, a third spotted, and a fourth brown. Among the bulls, the number of white ones is one half plus one third the number of the black greater than the brown; the number of the black, one quarter plus one fifth the number of the spotted greater than the brown; the number of the spotted, one sixth and one seventh the number of the white greater than the brown. Among the cows, the number of white ones is …

You get the picture. It's fantastically complicated. And the question at the end is: what's the composition of the herd?

Archimedes didn't go out of his way to encourage a solution, commenting that anyone who cracked the problem would be 'not unknowing nor unskilled in numbers, but still not yet to be numbered among the wise'. A couple of millennia later, German mathematician A. Amthor was able partially to claim this weak accolade by showing that the answer contained 206,545 digits and began with 7766. But, being shackled by having only log tables instead of a high-performance laptop, he gave up at that point. Finally, in 1965, mathematicians at the University of Waterloo, Canada, ran an IBM 7040 computer for seven and a half hours to arrive at the exact answer. Unfortunately, the 42-page printout subsequently went missing and it wasn't until 1981 that Harry Nelson, a colleague of one of the authors (David) at Cray Research, reran the calculation on a Cray-1 supercomputer. This time it took only 10 minutes to churn out the solution, which was condensed onto 12 pages of printout and then reproduced on a single page of the *Journal of Recreational Mathematics*.

Two of the greatest mathematical puzzlers were the American Sam Loyd and the Englishman Henry Dudeney, who both lived in the second half of the nineteenth century and into the early part of the twentieth. Loyd's genius for inventing tantalising and popular problems was matched only by his talent for self-promotion and downright deception. Among his best-known creations were the Hoop-Snake Puzzle, Get Off the Earth, and, most famous of all, the Fifteen Puzzle. At 17 he was already being hailed as a leading writer of chess problems before going on to become one of the strongest players in the US with a world ranking of 15. Also as a teenager he crafted the deceptively simple-looking Trick Mules Puzzle. The object of this was to cut apart three pieces that show two mules and two jockeys and then reassemble the pieces so that the jockeys are riding the mules. Loyd sold the puzzle to the showman Phineas T. Barnum (of Barnum & Bailey Circus fame) for around $10,000. Problems like this, which look so simple to solve that people felt compelled to try them only to find, hours later, that they still hadn't figured them out, became Loyd's forte. But he was often less than honest about the source of some of the puzzles he claimed to have invented.

One such conundrum, the Fifteen Puzzle, which Loyd said he'd come up with in the 1870s, became a worldwide obsession, much as Rubik's Cube did a century later. The object is to put fifteen tiles, numbered 1 to 15, which are held in a four by four frame, in serial order starting from a random position. One square is left empty and the only allowed moves consist of sliding tiles into the empty space. Everyone it seemed was caught up with the craze – playing the game in horse-drawn trams, during their lunch breaks, or when they were supposed to be working. The game even made its way into the solemn halls of the German parliament. 'I can still visualise quite clearly the grey-haired people in the Reichstag intent on a square small box in their hands,' recalled the geographer and mathematician Sigmund Günter who was a deputy during the puzzle epidemic. 'In Paris the puzzle flourished in the open air, in the boulevards, and proliferated speedily from the capital all over the provinces,' wrote a contemporary French author. 'There was hardly one country cottage where this spider hadn't made its nest lying in wait for a victim to flounder in its web.'

Loyd took credit for it in his *Cyclopedia of Puzzles* (published in 1914): 'The older inhabitants of Puzzleland will remember how in the early seventies I drove the entire world crazy over a little box of movable pieces which became known as the "14–15 Puzzle".' In fact the real inventor was another American, Noyes Chapman, the Postmaster of Canastota, New York. Loyd did, however, offer a $1,000 reward for the first correct solution to a variation on the game in which the starting position had the tiles in numerical order except that the 14 and 15 were swapped around. Many people tried to claim the prize but none could reproduce their supposed winning series of moves under close scrutiny. There's a simple reason for this, which is also the reason that Loyd was unable to obtain a US patent for the game. According to regulations, Loyd had to submit a working model so that a prototype batch could be manufactured from it. Having shown the game to a patent official, he was asked if it were solvable. 'No,' he replied. 'It's mathematically impossible.' Upon which the official reasoned there could be no working model and thus no patent!

When the Fifteen Puzzle is thoroughly analysed, it turns out there are more than 20 billion different starting arrangements of the tiles but these fall into just two groups. In one, all the tiles can eventually be manoeuvred into ascending numerical order, whereas in the other, tiles 14 and 15 will always be the wrong way round no matter what you do. It's impossible to combine arrangements from the two groups and impossible to turn an arrangement in one group into an arrangement in the other. Given a random setting of tiles, can you know in advance if you have one of the unsolvable kind? Yes: simply count how many times a tile numbered n appears after a tile numbered $n + 1$. If there's an even number of such inversions, the puzzle is solvable, otherwise you're wasting your time!

It was only in the 1890s that Loyd became a full-time professional puzzle-maker. Around the same time he struck up a regular correspondence with his counterpart across the Atlantic, writer and master puzzlist Henry Dudeney. Though he left school at thirteen to start work as a clerk in the civil service, Dudeney became expert at devising chess and maths problems. These he'd often describe in articles for newspapers and magazines, writing under the pseudonym 'Sphinx'. For thirty years he was a columnist for *Strand Magazine* and also wrote six books. The first of these, *The Canterbury Puzzles*, published in 1907, purports to include a collection of problems posed by the characters in Chaucer's *The Canterbury Tales*. One of the problems is called the Haberdasher's Puzzle and its solution is Dudeney's best-known geometrical discovery. The puzzle is to cut an equilateral triangle into four pieces that can be rearranged to make a square. A remarkable feature of the solution is that each of the pieces can be hinged at one vertex, forming a chain that can be folded into the square or the original triangle. Two of the hinges bisect sides of the triangle, while the third hinge and the corner of the large piece on the base cuts the base in the approximate ratio 0.982: 2: 1.018. Dudeney showed just such a model of the solution, made of polished mahogany with brass hinges, at a meeting of the Royal Society on 17 May 1905.

For a while Loyd and Dudeney collaborated in devising new puzzles, Dudeney being, by general reckoning, the superior mathematician and Loyd the more ingenious in his presentation and advertising skills. But in time a rift developed between the two on account of Loyd's tendency to borrow ideas without acknowledgement and even to publish them as if they were his own. One of Dudeney's daughters 'recalled her father raging and seething with anger to such an extent that she was very frightened and, thereafter, equated Sam Loyd with the devil'.

WEIRDER MATHS

Loyd also took credit for a vanishing puzzle called Get Off the Earth – one of his most commercially successful products, though it was based on similar and earlier designs. In a vanishing puzzle the total area of a collection of pieces, or the number of items in a picture, appears to change following some manipulation. In Get Off the Earth, published in 1896, the picture is made from a rectangular background of card on top of which a circular card, representing the world, can be rotated. Parts of a number of men, supposed to be Chinese, are on each piece. With the world orientated so that the large arrow on it points to the N.E. point on the background, 13 'Chinamen' can be counted. But when the Earth is turned slightly, so that the arrow points N.W., there are only 12 characters. The puzzle is: where did the 13th Chinaman go? The cleverness of the puzzle is that there are many bits of Chinamen – arms, legs, bodies, heads, and swords – and each has tiny slivers missing. When the Earth is turned, these pieces get slightly rearranged. In particular, each of the 12 Chinamen gains a sliver of a Chinaman from his neighbour.

Get Off the Earth is a form of optical illusion in that it tricks our senses. Other mathematical puzzles seem to defy our intuition about what's reasonable – but only because our intuition is often a poor guide. A simple example of this is when we're asked about how different quantities vary in one, two, and three dimensions. Imagine, for instance, that the Earth, taken to be a perfect sphere with a radius of 6,378 kilometres, is completely covered by a thin membrane. Now suppose that one square metre is added to the area of this membrane to form a larger sphere. By how much does the radius and the volume of the membrane increase? The answers are easy to work out from the formulae for the area and the volume of a sphere, respectively. Surprisingly, it turns out that if the area of the cover is increased by one square metre, then the volume it contains is increased by about 3.25 million cubic metres, a seemingly huge amount. However, the new cover wouldn't be very high above the surface of the planet – only about 6 billionths of a metre!

Another problem to do with spheres not only has an answer that's counterintuitive but it also seems, at first, to lack enough information to even get at the solution. Say you have a wooden bead that's one inch (about 2.5 centimetres) high and you drill a hole exactly through the middle of it so that the remaining part of the sphere is only half an inch high. Now imagine you had an enormously large drill and used it to bore a hole though the Earth so big that the part of the Earth left behind was only half an inch high. Amazingly, the residual volumes of these two holey spheres, the drilled bead and the drilled Earth, would be exactly the same! It just happens that even though the Earth is vastly larger than the bead, the drill would take out proportionately more in order to make the height of the hole the same, so that the volume left doesn't depend separately on the initial size of the sphere or of the hole, but only on their relation, which is determined by requiring the hole to be half an inch long. This fact enables the following poem-problem, which appeared in Louis Graham's book *Surprise Attack in Mathematical Problems*, to have a solution even though it seems as if the reader has been short-changed with regard to data:

Old Boniface he took his cheer,
Then he bored a hole through a solid sphere,
Clear through the center, straight and strong,
And the hole was just six inches long.

Now tell me, when the end was gained,
What volume in the sphere remained?
Sounds like I haven't told enough,
But I have, and the answer isn't tough.

Having already given away the secret that the volume that remains of a drilled sphere doesn't depend on the initial size of the sphere, we can cheat and give a kind of meta-argument that's much shorter than the geometric proof that starts from scratch. The volume left behind of any sphere with a six-inch-long hole through it must be the same as the volume left behind of a six-inch-diameter sphere with a hole of zero diameter drilled through it. The answer then follows immediately from the formula for the volume of a sphere ($V = 4/3 \ \pi r^3$) when the radius, r, equals 3 inches – approximately 113 cubic inches.

Any recreational maths problem, whatever its nature, has to fit the bill on two counts. First it has to be solvable without resort to anything beyond what most people are taught in school. And second, there should be an appealing or intriguing aspect to it that draws us in. Puzzle-makers have often found that this second aspect can be enhanced by weaving in a little faux history. French mathematician Édouard Lucas, remembered for his study of the Fibonacci sequence and discovery of a related sequence that now bears his name, used a bit of invented fiction to spice up a popular game he invented called the Tower of Hanoi. Early versions of it, first sold as a toy in 1883, came bearing the creator's name 'Prof. Claus' of the College of 'Li-Sou-Stain', which was quickly discovered to be an anagram of 'Prof. Lucas of the College of Saint Louis'. The game consisted of three pegs, on one of which were eight discs, stacked largest to smallest. The problem was to transfer the tower to either of the vacant pegs in the fewest possible moves, by moving one disc at a time and never placing any disc on top of a smaller one.

Lucas added some exotic appeal to his product with a romantic tale that accompanied the game. The Tower of Hanoi, he wrote, was a miniature version of the magnificent fabled Tower of Brahma. According to this myth, in the Indian city of Benares, beneath a dome that marked the centre of the world, is to be found a brass plate. On the plate are set three diamond needles, 'each a cubit high and as thick as the body of a bee'. The god Brahma placed 64 discs of pure gold on one of these needles at the time of Creation. Each disc is a different size, and each is placed so that it rests on top of another disc of greater size, with the largest resting on the brass plate at the bottom and the smallest at the top. Within the temple are priests whose job it is to transfer all the gold discs from their original needle to one of the others, without ever moving more than one disc at a time. No priest can ever place any disc on top of a smaller one, or anywhere else except on one of the needles. When the task is done, and all 64 discs have been successfully transferred to another needle, 'tower, temple, and Brahmins alike will crumble into dust, and with a thunder-clap the world will vanish'. The prediction seems fairly safe given that the number of steps required to transfer all the discs would be $2^{64} - 1$, or approximately 1.8447×10^{19}. Assuming one second per move, this would take about five times longer than the current age of the universe! Fortunately, the Tower of Hanoi doesn't have to be so time-consuming. With only eight discs to transfer, instead of 64, the minimum number of moves required is only $2^8 - 1$, or 255.

Like a number of other mathematicians, Lucas died in unusual circumstances. At the banquet of the annual congress of the Association française pour l'avancement des sciences, a waiter dropped some crockery and a piece of broken plate cut Lucas on the cheek. He died a few days later, at the age of only 49, of a severe skin inflammation probably caused by septicaemia.

Much older, but mathematically linked to the Tower of Hanoi, is another mechanical puzzle – one that involves moving things around – called Chinese rings, among other names. The object is to remove a number of rings from a horizontal loop of stiff wire and then put them back on. On the first move it's possible to take up to two rings off the left end of the wire. One or both of these can then be slipped through the wire loop, from top to bottom. If both are removed then the fourth ring can be slipped over the end. If just one of the first two is removed, then the next step is to slip the third ring over the end. Subsequently, rings must be put back on to the wire loop in order to remove other rings, and this process is repeated over and over again.

In general, if n is the number of rings, the minimum number of moves needed is $(2^{n+1} - 2)/3$ if n is even and $(2^{n+1} - 1)/3$ if n is odd. Removing seven rings, for instance, can be done in 85 moves. Most of the solution is easy, as each move normally involves going forward or backward to the previous state. The key to a correct solution is the first step: if n is even, you must remove two rings; if n is odd, you must remove only one. It's a process similar to that involved in the Tower of Hanoi and, in fact, Édouard Lucas gave an elegant solution that uses binary arithmetic.

As with so many old mathematical pastimes, the origins of the Chinese rings are shrouded in mystery. According to nineteenth-century ethnologist Stewart Culin the puzzle was invented by a Chinese general, Zhuge Liang, in the second century CE, as a present to his wife so that she'd have something to occupy her mind while he was away fighting. One of the earliest references to it in Europe is in the manuscript *De Viribus Quantitatis*, written in about 1500 by the Italian mathematician and Franciscan monk Luca Pacioli. Problem 107 of this work is accompanied by the description: 'Do cavare et mettere una strenghetta salda in al quanti anelli saldi, difficil caso' (Remove and put a little bar joined in some joined rings, difficult case). It was also mentioned by another Italian, Girolamo Cardano, who was the first European to write about negative numbers. In the 1550 edition of his book *De Subtililate*, Cardano talks at length about the puzzle, which is why it's also sometimes referred to as Cardan's rings. By the end of the seventeenth century, the rings had become popular in many European countries. French peasants even used it as a means to lock chests and called it baguenaudier, or 'time-waster'.

Quite why so many maths puzzles are said to have come from China isn't clear. Perhaps, in years gone by, a purported Far East origin added an element of mystery and the exotic to a pastime that, in reality, had more humble roots. The game of Nim, of which there are many different versions, certainly resembles a Chinese game called *jiǎn-shízi* ('picking stones'), though the name 'Nim' was coined by Charles Bouton, an associate professor of mathematics at Harvard at the turn of the twentieth century. He took it from an archaic English word meaning to steal or take away, and in 1901 published a full analysis of Nim including a proof of the winning strategy. Nim involves two players alternately removing at least one item from one of two or more piles or rows. The person who picks up the last item wins. In one form of the game, five rows of matches are laid out in such a way that there is one match in the first row, two matches in the second, and so on, down to five matches in the bottom row. Players take turns to remove any non-zero number of matches from any one row.

The first Nim-playing computer, the Nimatron, a one-ton behemoth, was built in 1940 by the Westinghouse Electrical Corporation and exhibited at the New York World's Fair. It played 100,000 games against spectators and attendants and won an impressive 90 percent of the time. Most of its losses came at the hands of attendants who were instructed to reassure incredulous onlookers that the machine could be beaten! In 1951 a Nim-playing robot, the Nimrod, was shown at the Festival of Britain, and later at the Berlin trade fair. It was so popular that spectators entirely ignored a bar at the other end of the room where free drinks were being offered. Eventually the local police had to be called in to control the crowds.

In terms of difficulty, perhaps one of the toughest challenges in recreational maths ever put to the general public was the Eternity Puzzle. It was a jigsaw consisting of 209 pieces, each one different and each made from a unique configuration of equilateral triangles and half-triangles with the same total area as six triangles. The object was to fit them together into an almost-regular 12-sided figure aligned to a triangular grid. The puzzle's inventor, Christopher Monckton, announced a prize of $1 million when the puzzle was released commercially in June 1999, to go to the first correct solution submitted, assuming there was one, when all the solutions were opened in September 2000. Monckton had run computer searches on much smaller versions of the puzzle, which had convinced him that the sheer size of Eternity would make it intractable. However, the prize was won by two British mathematicians, Alex Selby and Oliver Riordan, with the help of a couple of computers, who sent in a correct tiling on 15 May, six weeks ahead of the only other puzzler known to have found a correct solution.

Early on, Selby and Riordan made a surprising discovery. As the number of pieces in an Eternity-like puzzle increased, so did the difficulty – but only up to a point. The critical size is about 70 pieces, which would be almost impossible to solve. For larger puzzles, however, the number of possible correct solutions increases. In the case of Eternity itself, with its 209 pieces, there are thought to be at least 10^{95} solutions – far more than the number of subatomic particles in the universe but far, far less than the number of non-solutions. The puzzle itself is much too large to solve by an exhaustive search but not, as it turns out, by more savvy methods that take into account what shaped regions are easiest to tile and what shaped pieces are easiest to fit. By steadily refining their search algorithm, Selby and Riordan were able to prune out the vast majority of non-solutions and, with a bit of good fortune, to hit upon a correct solution and claim the prize.

While many maths puzzles are intended just for fun, a few, although easily stated, have led to groundbreaking developments. The most famous of these, which we've talked about elsewhere, is the Bridges of Königsberg. Leonhard Euler's resolution of the problem – in the negative – marked the birth of graph theory and was an important early step in topology. Another long-standing conundrum, the Four-Colour Problem, was to prove – or disprove – the assertion that you need no more than four different colours to colour any map so that no two adjacent regions are coloured the same. It's easy to show the assertion's true for specific cases but fiendishly hard to come up with a proof that covers every possibility. A proof was finally announced in 1976, by Kenneth Appel and Wolfgang Haken at the University of Illinois, and marked the first time that a computer had played a major role in such an achievement. A far-reaching generalisation of the Four-Colour Problem was proposed by the Swiss mathematician Hugo Hadwiger in 1943 and remains one of the greatest unsolved problems in graph theory.

It was while contemplating the Four-Colour Problem that Piet Hein, in 1942, hit upon the idea for a new board game, which became popular in Denmark under the name *Polygon*. Independently, the same idea came, a few years later, to American John Nash, leading game theorist and subject of the biography and film *A Beautiful Mind*. Nash's version was played by maths students at Princeton and a number of other American campuses and eventually marketed by Parker Brothers as Hex – a name that stuck. Hex featured in Martin Gardner's *Mathematical Games* column in July 1957 and has been the subject of numerous studies in game theory. Nash himself was the first to prove that Hex can't end in a draw and that the first player, no matter what the board size, always has a winning strategy.

Whether it's playing Hex, chess, mancala, or tic-tac-toe, creating a string figure in Cat's cradle, solving a maze or a logic puzzle, folding a flexagon or an origami model, or weaving a braid, we're doing something mathematical. Just as art and music come in many forms so, too, does maths. Far from being a dry and difficult subject as it's sometimes portrayed, maths can be a delightfully playful and human thing – something that we do, often without knowing, just for the fun of it.

CHAPTER 12

Shapes Weird and Wonderful

Strangeness is a necessary ingredient in beauty.

– Charles Baudelaire

AMERICAN AIRMAN JOSEPH Portney was aboard a US Air Force KC-135 flying over the North Pole in 1968 as an engineer checking out some new navigational equipment. As he looked down on the icy terrain he had a strange thought: what if the Earth were a different shape? What if its oceans, continents, islands, and polar caps were mapped onto a cylinder, a pyramid, a cone, or a torus? When he got home, he sketched and captioned 12 different hypothetical Earths and gave them to the graphic arts group at the company where he worked, Litton Guidance & Control Systems, to create models. These were then photographed and became the theme of a Litton publication called *Pilots and Navigators Calendar for 1969*. Each month was introduced with a different one of the 12 hypothetical Earths. The result was an international sensation, attracting awards and heavy fan mail.

Portney's fascination with shapes and geometries has been shared by countless others over the centuries. What's been uncovered is an astonishing zoo of different curves, surfaces, solids, and higher-dimensional forms, only some of which can be made into real objects. The rest, for one reason or another, can't exist in this world but only in the weird realm of thought that's home to everything mathematically possible.

Some shapes aren't hard to imagine or depict yet still have seemingly bizarre properties. One of these goes by the name Gabriel's Horn or Torricelli's Trumpet, because it was first investigated by the Italian physicist and mathematician Evangelista Torricelli in the early seventeenth century. As a young man, Torricelli studied in Galileo's home at Arcetri, near Florence, and then, upon Galileo's death, succeeded his teacher as mathematician and philosopher for their mutual friend and patron, the Grand Duke of Tuscany. Although best known for his invention of the barometer he also made important contributions to maths, none more so than his discovery of the famous Horn, which prompted a fierce debate about the nature of infinity while delighting others. Torricelli's mathematical compatriot Bonaventure Cavalieri wrote:

> I received your letter while in bed with fever and gout … But in spite of my illness, I enjoyed the savoury fruits of your mind, since I found infinitely admirable that infinitely long hyperbolic solid which is equal to a body finite in all the three dimensions. And having spoken about it to some of my philosophy students, they agreed that it was truly marvellous and extraordinary that that could be.

Gabriel's Horn is the surface of revolution of the curve of $y = 1/x$, a rectangular hyperbola, for values of x greater

Gabriel's Horn.

than 1. A surface of revolution is just a surface produced by rotating a line or a curve about some axis. For example, a sphere is the surface of revolution that comes from spinning a circle about its diameter. Gabriel's Horn is formed when the curve of $y = 1/x$, for $x > 1$, is rotated about the x-axis. Torricelli was astonished to discover that although the Horn has a finite volume, equal to π cubic units, it has an infinitely large surface area! How could a surface with an infinite area enclose a finite volume? Torricelli tried, by various methods, to prove that the area was in fact finite, but failed.

This very odd state of affairs led to what's been called the 'painter's paradox' because it seems that if Gabriel's Horn were filled with paint there wouldn't be enough of it even to coat the surface. Surely you can't cover something that's infinitely big with something else of which there's only a finite amount. Yet if you fill Gabriel's Horn with paint, there must surely be enough of it (with plenty left over) to cover its entire inner surface. Using real paint, made of atoms and molecules, this is true enough. There comes a point at which the Horn becomes too narrow for any more molecules of paint to fit, so the paint ends up covering only a finite part of the surface. Also, if we assume atoms to be spheres, they only touch the surface at single points, so what it means for the paint to 'cover' the surface becomes less clear. The fact is that if we're talking about a real-world situation in which physical paint's involved we have to make the Horn physical too. Most significantly, this means that it narrows off to a certain point at which it's so thin that it's less than an atom or molecule wide. A physical Horn must end here, resulting in both its volume and surface area being finite.

A true (mathematical) Horn is the thing that perplexed Torricelli. As word of his discovery spread, others marvelled at and wondered what it meant. It suggested, more than ever, that there might be such a thing as 'completed' or 'actual' infinity – in this case, a shape that was truly infinitely long – as distinct from 'potential infinity', or simply going on indefinitely forever. The English philosopher Thomas Hobbes was among those who had a fair bit to say about the Horn, especially because it didn't chime well with what he thought infinity should be like.

With the benefit of hindsight, we can appreciate that using special, mathematical paint, which can be applied as thinly as need be, without limit, the paradox that tasked mathematicians and philosophers at the time of Torricelli never arises. The thickness of the coat can be made vanishingly small at a rate quickly enough to compensate for the ever-expanding area, enabling a strictly finite volume of paint to coat an infinitely large surface. Unfortunately, Torricelli lived just before calculus appeared on the scene. Otherwise, he would have understood that the apparent paradox of the Horn could be explained in terms of the infinitely small quantities known as infinitesimals.

Gabriel's Horn is also interesting in that it has negative curvature. This puts it in the same category as some other intriguing surfaces such as the pseudosphere. As the name, meaning 'false sphere', suggests, the pseudosphere and the sphere are closely connected. What separates them is the nature of their curvature. A sphere has positive curvature at every point. In other words, its surface always stays on one side of a plane, known as a tangent plane, that just touches the surface at any point. In contrast, a surface has negative curvature at a point if the surface curves away from the tangent plane in two different directions. Not only does a sphere have positive curvature everywhere but it also has *constant* positive curvature, equal in value to $+1/r$, where r is the sphere's radius. The pseudosphere is exactly the opposite, having everywhere a constant negative curvature, equal to $-1/r$. For a given value of r, the sphere and pseudosphere enclose the same volume. However, whereas a sphere has a closed surface and a finite area, a pseudosphere has an open surface and a finite area. (In the matter of area, the pseudosphere differs from Gabriel's Horn because it narrows off more quickly.) Another result of the pseudosphere's negative

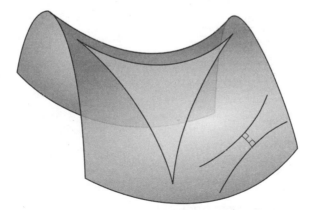

On the surface of a shape with negative curvature the angles of a triangle add up to less than 180° and lines that start off parallel diverge.

curvature is that the angles of a triangle drawn on its surface add up to less than 180°; on a sphere, a triangle's angles sum to more than 180°.

The geometry on the surfaces of both the sphere and the pseudosphere doesn't follow the rules prescribed by Euclid, which apply only to a flat plane. Both are examples of non-Euclidean geometry: spherical (or elliptical) geometry in the case of the sphere and hyperbolic geometry in the case of the pseudosphere. Scientists, since the time of Albert Einstein, have been aware of the fact that the space in which we live is curved by the presence of what it contains, namely matter and energy. What remains uncertain, however, is the overall shape of the universe, a fact that is governed by the average density of matter and energy that it contains. It could be a shape that's analogous to a sphere, a pseudosphere, or a flat plane. The best observational data available at present suggests that the universe is almost exactly flat, which, if true, means that it will expand forever.

In the same way that Gabriel's Horn is the surface of rotation of part of the curve $y = 1/x$, so the pseudosphere comes from rotating a curve known as the tractrix about the axis which it approaches more and more closely without ever touching. The tractrix is the answer to a question asked by Frenchman Claude Perrault. Not known as a giant of mathematics, Perrault trained as a doctor and gained a minor reputation as an architect and an anatomist before dying in unusual style as a result of an infection he caught while dissecting a camel. His greatest claim to fame, aside from his connection with the tractrix, is that he was the brother of the author of *Cinderella* and *Puss in Boots*. In 1676, at about the time German mathematician and polymath Gottfried Leibniz was doing groundbreaking work on the calculus,

Perrault placed his pocket-watch on the middle of a table, pulled the end of its chain along the edge of the table, and asked: what is the shape of the curve traced by the watch?

The first known correct answer to Perrault's question came in a letter to a friend in 1693 by Dutch physicist, astronomer, and mathematician Christiaan Huygens, who also coined the name 'tractrix' from the Latin *tractus* for something that is pulled along. (The corresponding German name is *hundkurve*, or 'hound curve', which makes sense if you imagine the path a dog might follow on its leash as its master walks away.)

The tractrix is closely related to another interesting curve known as the catenary, which is the shape taken up by a free-hanging chain. In fact, the name comes from the Latin *catena* for 'chain'. Power cables suspended between pylons also form a catenary, as does the path of a moving charge in a uniform electric field. A tractrix can be drawn starting from a catenary in a very simple way. Imagine you've fixed a piece of string at one end to a point on the catenary. Pull the string out so that it forms a tangent to the curve where it's attached. Then wind up the string being careful always to keep it taut. The path followed by the end of the string will be a tractrix. If you were to do the same thing starting with a circle, the result would be a type of spiral. (Or think of the path that a goat, tethered to a post, would follow if it walked around and around in the same direction, keeping its tether taut until it wound its way to the centre.) In both cases, the shape obtained is known as the involute of the original curve.

Rotate a catenary about its central axis and yet another fascinating shape, the catenoid, emerges. First described by Swiss mathematician Leonhard Euler in 1740, it's the oldest known minimal surface – a shape of least area when bounded by a given closed space – other than the plane. It's the only known minimal surface that's also a surface of revolution and is the minimal surface connecting two parallel circles of unequal diameter on the same axis. A way to make one, as we saw in Chapter 10, is to dip two circular rings into a soap solution and slowly draw the circles apart.

Of all surfaces of revolution none is more surprising than the superegg, which was named and popularised by Danish poet and scientist Piet Hein. A superegg emerges from the rotation of a certain kind of superellipse – a shape that's midway between an ordinary ellipse and a rounded rectangle. The equation of a run-of-the-mill ellipse is $(x/a)^2 + (y/b)^2 = 1$, where a is half the length of the ellipse's longer axis and b is the half-length of its shorter axis. In the nineteenth century, French mathematician Gabriel Lamé studied the family of curves produced by the more general equation $|x/a|^n + |y/b|^n = 1$, where the upright lines mean 'absolute value' (the unsigned value of whatever's between the lines) and n is bigger than 0. Not surprisingly, the family became known as Lamé curves. The ellipse is just the Lamé curve for which $n = 2$. A four-pointed star shape, called an astroid, is the result when $n = 2/3$. For all values of n bigger than 2, Lamé curves are known as superellipses. The superegg is the surface of revolution of the superellipse for which $n = 2.5$ and $a/b = 6/5$. Its strangeness becomes apparent only when it's made into a real, physical object formed, for example, out of wood. As Piet Hein pointed out, a superegg stood on either end has a peculiar and surprising stability, so much so that playing with one is a rather satisfying experience. In the 1960s, supereggs made of metal, wood, and other materials began to be sold as novelties and, in particular, small, solid-steel ones were marketed as executive toys. Today, from Piet

Hein's own website, you can order, complete with a tasteful grey leather bag, a stainless steel superegg, whose 'soft curves combined with the cold steel make it perfect for de-stressing play and fidgeting'. Or you may care to visit the world's largest superegg, made of steel and aluminium and weighing one ton, which was placed outside Kelvin Hall in Glasgow in 1971 to honour Hein's appearance as a speaker there.

The history of the superegg goes back to 1959 when city planners in Stockholm were looking to complete a redesign of Sergels torg (Sergel's Square), the most central of Stockholm's public squares. It was decided to finish the redesign with a fountain surrounding a monument, around which traffic would flow in a roundabout. For the shape of the fountain the chief designer of the project consulted his friend Piet Hein, who took less than a minute to come up with a 'continuously varying bending shape' – the superellipse generated by the equation we mentioned earlier. Later, the ingenious Hein spun his special superellipse into the solid which, sold as a popular novelty item, proved for him to be a golden egg.

The fame of the superellipse, however, didn't end with its use as a traffic island in the Swedish capital. It became the iconic shape of Scandinavian tables of that era and so of contemporary tables in general in the 1960s. When negotiators from opposing sides in the Vietnam War couldn't agree on the shape of a table for their meeting in Paris, they finally settled on the superellipse. On a grander scale it was chosen as the form of the main stadium for the 1968 Olympic Games in Mexico City.

It's easy to make a curved object that if put down on a flat surface always comes to rest in the same position – simply weight it at one end. The superegg is special, and entertaining, because it does the trick without any kind of built-in bias: it's made of material of the same density throughout. An even more astonishing stability is displayed by a shape that's been described as the strangest object in the world. It certainly has a strange name: the gömböc, from the Hungarian for 'sphere' (because it has some sphere-like properties). The existence of the gömböc was first conjectured in 1995 by Russian mathematician Vladimir Arnold. It's defined to be a convex 3D homogeneous (same consistency throughout) object that, on a flat surface, has just one stable and one unstable point of equilibrium. In other words, if placed on a flat surface, in every position except one it will move until it's resting on its solitary stable point; the only exception is the unstable point in which it will remain unless given the slightest nudge. Proof that the gömböc isn't some mythical beast but can exist in the real world, and in many different forms, came from Hungarian mathematician and engineer Gábor Domokos and his student Péter Várkonyi in 2006.

To look at, a gömböc isn't much. It has a broadly curved base surrounded by flattish sides that rise to a curved ridge. Put it down on the curved base and, as you might expect it rocks back and forth until it comes to rest on its one stable point. If laid down on a flat side it eventually reaches the same resting point but more slowly and in a way that almost suggests it's alive. A slow back-and-forth rolling motion is followed by a temporary pause before the gömböc goes into a rapid, rolling vibration, which ends with it reaching its stable balance point and righting itself again.

Gömböcs can be purchased today from many sources but they aren't cheap and their ability to self-right consistently depends much on how carefully they're made and their composition (heavier materials tend to be more effective).

The proportions of a handheld version, for example, must be accurate to within about one hundredth of a millimetre, or a tenth the thickness of a human hair, in order to work properly. *The New York Times Magazine* chose the gömböc as one of its seventy most interesting ideas in 2007 and further fame for the shape came a couple of years later when the gömböc appeared on BBC TV's popular quiz show *QI*. Host Stephen Fry demonstrated the curiosity before Gábor Domokos, who was in the audience, explained how it came about – and its connection with tortoises.

Tortoises are among the animals that can be in big trouble if they get tipped onto their backs, by accident or in a fight. The ability to self-right then becomes crucial for survival. Some tortoises and turtles, especially those with flattened shells, have long legs and necks with which they can lever themselves upright again. But those whose shells are more rounded need a different strategy. Following their breakthrough in tracking down the gömböc, Domokos and Várkonyi spent a year at the Budapest Zoo measuring and analysing the shapes of shells of different species of tortoise. Their resulting explanation of tortoise body shape and ability to self-right, in terms of gömböc geometry, while still controversial, has been accepted by some biologists.

Other shapes, some long known, achieve their stability through a combination of shape and spin. The most extraordinary of these is a shape that's been known for thousands of years and crops up with different names and in different cultures as diverse as Egyptian and Celtic. Known variously as a rattleback, a celt, and a wobblestone, it has a boat-shaped form with a curved base and a roughly elliptical top. Spun in one direction it will rotate a few times, then rattle from end to end, and finally reverse its direction of motion.

Spun the other way it will just continue in that direction before coming to a halt. Its surprising behaviour can be traced to the fact that the shape of the bottom is not completely symmetrical: it's higher on one side than the other.

Rotation is at the heart of another puzzle that, again, is of interest to both mathematicians and physicists or engineers. Suppose you need a two-dimensional shape that can be used as a roller. It must have a constant width while turning, otherwise anything rolling on top of it will bob up and down. Obviously, a circle will do the job – in fact, it may seem at first to be the only possible shape that would work. But, surprisingly, there are others. The simplest is the Reuleaux triangle, named after the German mechanical engineer Franz Reuleaux who developed machines that translated one type of motion into another. To make such a triangle, take an equilateral triangle and draw the circular arc centred at a vertex through the other two. Do this with all three vertices and you obtain a triangle with curved sides and a constant width. As rollers, Reuleaux triangles work just as well as circles (although they're not so good at being wheels, which require a constant radius as well as a constant diameter).

Reuleaux triangles, like any curves of constant width, make effective manhole covers – the key property here being that if the cover is moved or dislodged it isn't able (as, say, a square cover would be) to fall down the shaft. Their most ingenious application though is as a drill bit. Watts Brothers, a tool manufacturing company based in Pennsylvania, invented a drill bit, based on the Reuleaux triangle, which can drill (almost) square holes! The four sides of the hole are all perfectly straight, although, because the Reuleaux triangle has angles of 120°, it can't get completely into the corners of the triangle and so leaves them rounded.

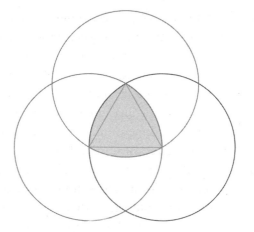

How a Reuleaux triangle (shaded area) can be made from an equilateral triangle and three circles.

Reuleaux triangles can be extended to other polygons. For example, a similar method of construction gives rise to Reuleaux pentagons and heptagons. The Reuleaux heptagon is particularly familiar in the United Kingdom – it's the shape of a British 20p or 50p coin. The reason for using a curve of constant width is that the coins will always be able to fit inside a vending machine regardless of which orientation they are entered in, and the unusual shape, as distinct from a circle, makes it much harder to counterfeit.

In three dimensions, a shape with strange rolling properties is the sphericon, discovered in 1979 by Israeli toy inventor David Hirsch. As with Reuleaux, Hirsch's goal was to come up with a device that produced a certain kind of motion – in this case, a waddling movement for a pull-along toy. In 1980 he filed a patent for his invention and the following year Playskool Company began marketing a toy called Wiggler Duck based on Hirsch's innovation.

To make a sphericon, start with a right circular cone – a cone with a circular base and an angle at the top of 90°. Now glue two of these together to form a double-cone. From a side-on view, because of the right angle at the top, it should have a square profile. Now, slice it in half vertically, in a plane that contains both vertices. This will result in two identical parts, each with a square cross section. The key step comes next: rotate one of the halves through 90° and then glue the two halves back together. Voila: you now have a sphericon.

It's a shape with some unusual properties. Unlike an ordinary cone or double-cone, which only rolls in circles, the sphericon can roll in a straight line – though, due to the conical shapes of the faces, the line is slightly wavy rather than perfectly straight. If you have two sphericons, they can roll on each other as well as they can on a flat surface. In fact, you can surround a sphericon with eight others, and all of them can roll simultaneously on the one central shape.

All of the shapes we've talked about so far, however strange, can actually be made or, at least, approximated. True, we can't build a complete model of Gabriel's Horn or a pseudosphere that stretches out to infinity. But we could make a finite model and then just imagine it stretching away forever. But there are some mathematical shapes that are so weird or outrageous in their properties that no physical representation can capture their essence. Among these mathematical miscreants are so-called pathological shapes, the properties of which can often defy intuition. Weirdest of all is a structure known as Alexander's horned sphere.

Named after the Princeton mathematician James Alexander, who first described it in the early 1920s, the horned sphere is an example of what, in topology, is described as a 'wild' structure. On the inside, the horned sphere, in the eyes of topologists at least, is no different from a sphere. What this means is that it's simply connected and can be deformed into an ordinary sphere, without breaks or tears. On the outside, however, it's an entirely different story. This is where the horns come in. Externally, the horned sphere is made of an infinitely recursive set of interlocking pairs of rings of decreasing radius – horns within horns within horns, and so on, forever. Despite having a single spherical inside, this bizarre shape has an infinitely complex exterior. A rubber band placed around the base of any horn couldn't be removed from the structure even after infinitely many steps. Although the horned sphere is impossible to make, the American sculptor Gideon Weisz has modelled a number of approximations to the structure.

In the philosophy of Plato, one of the greatest Greek thinkers, four of the solids named after him are associated with the four classical elements: the cube with earth, the octahedron with air, the tetrahedron with fire, and the icosahedron with water. The fifth Platonic solid, the dodecahedron, was more loosely linked to the element of the heavens, known variously as aether or quintessence. Much later, Johannes Kepler matched the same five Platonic solids with the five then-known extraterrestrial planets. Today, our scientific worldview is vastly more sophisticated. Yet still it has room for speculation about deep connections between geometric shapes and fundamental physics. Astonishingly, theoreticians have recently found that the volume of a multidimensional object known as the amplituhedron, which resembles a multifaceted jewel, yields the solution to a complex series of formulae describing elementary particle interactions. Solving these formulae in the usual way can, in some situations, be too time-consuming even using high-speed computers. Yet calculations based on the amplituhedron can be done simply with pen and paper. According to one of the physicists who developed the new idea, Jacob Bourjaily at Harvard University, 'The degree of efficiency is mind-boggling.'

The amplituhedron, or some similar crystal-like object, may even hold the key to understanding one of the great mysteries of science: how gravity and quantum physics can be reconciled. Not only does the new geometric approach to particle interactions simplify the maths but it also suggests that we need to change the way we think about the nature of things. In the amplituhedron scheme, space, time, and the motion of particles in the arena of space–time are seen to be illusions. Instead of change – particles colliding and scattering, and forces acting over distance and time – what matters is the timeless structure of certain forms. Physical reality emerges, according to this startling new view, from the weird and wonderful structure of shapes whose existence we're just now beginning to discern.

CHAPTER 13

The Great Unknowns

> The mistakes and unresolved difficulties of the past in mathematics have always been the opportunities of its future.
>
> – E. T. Bell

THE LURE OF the unsolved is the lifeblood of mathematics. Everyone loves a good mystery and most of us enjoy brainteasers – Sudoku, logic puzzles, mazes, and the like. Mathematicians are no different, except perhaps that their curiosity runs deeper. The challenge of venturing into the uncharted realms of curious numbers, exotic geometries, or abstract algebras can prove to be a powerful intellectual aphrodisiac. And there's no danger of the quest to penetrate the great mathematical unknown ever reaching an end. Very often, the solution to one problem raises new ones and may even open up entirely new branches of maths.

The ancient Greeks were especially fond of geometry and particularly obsessed with three geometric riddles. All three of these unsolved problems had to do with construction, where only a straightedge and a compass are allowed. A surprising amount is possible with just these two simple tools, such as dividing a line segment into any whole-number ratio and constructing various regular polygons. Of the latter, equilateral triangles and squares proved the easiest, but the Greeks could also construct regular pentagons, 15-sided polygons, and polygons with double the sides of the basic constructible ones. So, from a 15-sided polygon, the Greeks could, for instance, develop a 30- or 60-sided regular polygon using only a straightedge and compass. However, three problems in construction obstinately resisted every form of attack.

Angle trisection was one of these outstanding conundrums. Given an arbitrary angle, is it possible, using just a straightedge and compass, to divide it into three equal parts? Some angles – such as a right angle (90°) – are easy to trisect in this way. But special cases aside, the Greeks found angle trisection stubbornly resistant to all their efforts. They could do it if allowed a *marked* ruler and a compass – a so-called neusis (from the Greek *neuein* 'incline towards') construction – but, with the odd exception, not with an unmarked straightedge.

The second geometry problem that stumped the Greeks was 'squaring the circle'. Given a circle, is it possible, again using just a straightedge and compass, to construct a square with the same area as the circle? In the fifth century BCE, Hippocrates of Chios seemed to make progress towards a solution by proving that a particular lune (a crescent shape with two circular arcs) has the same area as a triangle. His result showed that you can construct a triangle (and, with some more work, a square) with the same area as a shape with curved sides. But no one could extend this result to fully square the circle.

The third of the classic construction problems was 'duplicating the cube'. Given a cube, is it possible, with the two minimal tools, to make a cube with twice the volume? Again, the Greeks found that it was possible with a marked ruler but not otherwise. Two thousand years would pass before anyone took a step further and when the breakthrough came it was because of a new field of mathematics about which the Greeks knew nothing.

In 1796, while still a teenager, the great German mathematician Carl Gauss found a way to construct 17-sided polygons (and, by extension, those with multiples of 17 sides – 34, 51, 68, and so on). He was also able to say which polygons, including heptagons and nonagons, couldn't be produced by his technique. What's more, all three of the classic Greek problems proved resistant to Gauss's method. For a while, the possibility remained that other new approaches might yield the long sought-after prize of the Hellenic geometers. But after a few decades that hope was dashed forever, at the hands of a little-known French mathematician Pierre Wantzel, whose life was shortened by self-neglect.

After Wantzel died, a mathematical compatriot of his wrote: 'Ordinarily he worked evenings, not lying down until late; then he read, and took only a few hours of troubled sleep, making alternately wrong use of coffee and opium, and taking his meals at irregular hours …' His greatest claim to fame came in 1837, when he proved, once and for all, that trisecting the angle and duplicating the cube were impossible, and that Gauss's method could construct everything it was possible to construct using a straightedge and compass alone. There was absolutely no hope of any further breakthroughs on these matters.

Both Gauss's and Wantzel's attacks on the three classic problems of ancient geometry relied on a branch of maths pioneered by two Frenchmen, René Descartes and Pierre de Fermat, in the 1630s. Now known as analytic geometry, it starts from the premise that any point on a plane can be represented by two numbers, called Cartesian coordinates – the values of the point on the x and y axes. Historians see precursors to this field in Menaechmus of fourth century BCE Greece and the Persian mathematician, astronomer, and poet Omar Khayyam. But the flowering of the idea that geometry could be expressed using algebra had to wait, like so many other scientific revelations, until the dawn of the Renaissance in Western Europe.

A key aspect of analytical geometry is that certain distances can be represented as the roots of polynomials. A polynomial (from *poly–* meaning 'many' and *–nomial* meaning, in this case, 'term') is an expression like $4x + 1$, or $2x^2 - 3x - 5$, or $5x^3 + 6x - 1$. In other words, it's a combination of terms that includes constants (like 1 or -8), variables (like x or y), and exponents (like the 2 in x^2). A root of a polynomial is a value or values of the variable that makes the polynomial equal to zero. For example, the roots of the polynomial $x^2 + x - 2$ are 1 and -2, because if you put these numbers in place of x you'll zero out the expression. In analytical geometry, the problem of constructing polygons became the question of deciding which polynomials had roots corresponding to a distance constructible using straightedge and compass. Gauss found a way to construct all distances whose polynomials had a degree (the highest power of x appearing in the polynomial) equal to a power of 2. The 17-sided polygon resulted in a polynomial of degree 16, so it could be constructed.

Wantzel showed, using Gauss's method, that both angle trisection and cube duplication were impossible because they led to cubic polynomials – polynomials with degree 3.

His proof meant that mathematicians could stop looking for other possible ways of tackling these problems. There were simply none to be found, no matter what future developments came along, and no matter how many armchair theoreticians or cranks tried to convince others they'd come up with a solution.

That left only squaring the circle in limbo. For this construction to be possible using unmarked rule and compasses it would have to turn out that the number π (pi) – the ratio of the circumference to diameter of a circle – was the root of a polynomial of degree 2. This seemed highly unlikely, even as early as the seventeenth century. In 1882, all hope of ever being able to square the circle was lost when Ferdinand Lindemann proved that pi is transcendental – that is, a number which is not the root of any polynomial.

The Greeks' attempts to solve all three of their great construction problems, we now know, were doomed to failure from the outset. But it wasn't as if they'd overlooked something or taken a wrong turn. They simply didn't have the tools at the time to settle these issues, any more than they had the means to measure the distance to the nearest star or prove the existence of atoms. The fact is, it's often only in retrospect that we realise there was a critical gap in our knowledge that stopped a problem being solved. It may turn out to be a small gap, easily bridgeable, or a yawning gulf, similar in technological terms to that between the first birdmen of antiquity and the Apollo Moon landings. Interestingly, in the case of perhaps the most famous and, until recently, unsolved problem in maths it was probably the latter but, thanks to a little piece of centuries-old handwriting, discovered by chance, we can't be 100 percent sure that there isn't some simpler solution.

In 1637, while reading a copy of *Arithmetica*, by the Greek mathematician Diophantus, Pierre de Fermat wrote a note in the margin that would tantalise mathematicians for centuries. It was found, only by accident, twenty years after his death, by his son. In the note, Fermat claimed that the equation $a^n + b^n = c^n$ has no solutions where a, b, c, and n are positive integers and n is greater than 2. He went on to say that he had a proof of this, but that the margin of the book was too small to contain it. Did he really have a correct proof? Did he believe he had a proof but one that was actually wrong? Or was he joking, and perhaps pretending to know something that he didn't so that other mathematicians would be challenged to pursue the problem until it was finally resolved?

Fermat's Last Theorem, as it became known (although it was really just a conjecture), is easy to state and Fermat himself made an early breakthrough by proving and publishing it for the case of $n = 4$. Yet proving it for even one more value of n turned out to be surprisingly hard. Leonhard Euler eventually managed to do it for $n = 3$, in 1770, more than a century after Fermat. The Frenchmen Adrien-Marie Legendre and Peter Dirichlet supplied a proof for $n = 5$ in 1825, and a small army of other mathematicians, over the next century or so, tackled the problem from the point of view of specific exponents. Eventually, computers were brought into the fray to muscle the way to higher and higher powers of n. By early 1993, electronic number crunching had shown that Fermat's Last Theorem held for all values of n less than 4 million. By everyday standards that might seem like good grounds on which to assume the theorem is true in general. But mathematicians demand proof – rigorous, irrefutable, permanent proof that applies to all cases.

When it finally came, it was from a direction that no one had foreseen.

Between 1955 and 1957, two Japanese mathematicians, Yutaka Taniyama and Goro Shimura, came up with a suggestion that linked two seemingly very different areas of maths. One of these was elliptic curves, which are curves – confusingly *not* ellipses – described by a certain type of cubic equation. For example, the equation $y^2 = x^3 + 5x - 2$, if plotted out, would give an elliptic curve. The other area of maths involved in Taniyama and Shimura's work is known as modular forms. Think of a modular form as being a mathematical machine, with intricate internal parts like those of a pocket-watch, which can take an elliptic curve and assign to it a number. Because it bridged two such seemingly different mathematical territories, the so-called Taniyama–Shimura Conjecture was seen to have deep significance, at least by those who understood it. But it attracted much more attention after German mathematician Gerhard Frey, in 1986, suggested that a proof of it would also imply a proof of Fermat's Last Theorem. There was only one catch – proving the Taniyama–Shimura Conjecture looked fiendishly hard or, even, some mathematicians thought, impossible. Seven years later, such pessimism evaporated when British mathematician Andrew Wiles supplied a proof, although it took him an extra eighteen months to fix a serious error that he'd overlooked. Overnight, Wiles became a mathematical celebrity, made worldwide headlines (something almost unheard of for a mathematician), and was subsequently knighted. He'll always be remembered as the man who finally vindicated Fermat, although, in fact, his far greater achievement was proving a case of the Taniyama–Shimura Conjecture, which quickly led to a complete proof of this much more profound assertion.

Goro Shimura lived to see the work for which he's now best known become part of mathematics for eternity. Not so his colleague Taniyama, who originated the idea for the conjecture that lay at the heart of Wiles's great thesis. We saw in Chapter 8 how the intensity of mathematical work can sometimes, in those already vulnerable, lead to tragic outcomes. In November 1958, while serving as an assistant at the University of Tokyo and engaged to be married, and having been invited to join the prestigious Institute for Advanced Study, in Princeton, Taniyama took his own life at the age of 31. In his suicide note he wrote:

> Until yesterday I had no definite intention of killing myself. But more than a few must have noticed that lately I have been tired both physically and mentally. As to the cause of my suicide, I don't quite understand it myself, but it is not the result of a particular incident, nor of a specific matter. Merely may I say, I am in the frame of mind that I lost confidence in my future.

To compound the tragedy, a month later, his fiancée Misako Suzuki also committed suicide.

Fermat couldn't have come up with Wiles's proof 400 years earlier, any more than Galileo could have pioneered quantum mechanics. It also seems wildly improbable that he'd found a simpler proof, given that some of the best mathematical minds of the next four centuries tried and failed using any methods that he would have had available. He wasn't the calibre of mathematician either who'd fail to spot a mistake in some straightforward line of reasoning. So, the likeliest scenario remains: that he was simply playing a prank – perhaps with the intention of provoking others to look deeper into the problem. If so, it did the trick.

As long-standing problems are solved and mathematicians push deeper into previously uncharted realms of their subject, it's natural to ask if at some point they'll know everything there is to know. In the second half of the nineteenth century, some scientists started to believe that all the fundamental workings of nature could be understood in terms of Newtonian mechanics and Maxwellian electromagnetism. German physicist Philipp von Jolly, in 1878, even advised one of his students not to go into physics because 'in this field, almost everything is already discovered, and all that remains is to fill a few unimportant holes.' Fortunately, the student in question went into theoretical physics anyway – his name: Max Planck.

In the early part of the twentieth century there was a similar feeling in mathematics that perhaps some endgame was in sight. David Hilbert, the great German mathematician, proposed a project to show that all of maths would inevitably flow from a correctly chosen set of axioms – basic rules and statements that are assumed to be true at the outset. Prior to this, in 1900, he'd published a list of 23 unsolved problems, which included the axiomatisation of arithmetic as one of its challenges. Hilbert's list is generally recognised to be the most carefully constructed and influential ever compiled in the subject by an individual. It certainly acted as the stimulus for an immense body of work by subsequent generations of researchers.

Of the 23 problems, 10 are now considered to have been fully resolved. A further seven have been either partially resolved or have reached a state where the outcome rests on what starting assumptions are made. Into this last category fall the first two of Hilbert's problems, both of which touch upon questions of infinity and the foundations of the system of mathematics we choose to use.

Back in the 1870s, the German mathematician Georg Cantor showed that infinity comes in different sizes. In particular, he proved the startling fact that the infinity of natural numbers (1, 2, 3, and so on) is smaller than the infinity of real numbers (all the numbers representing points on a line). Cantor believed that there were no infinities intermediate in size between these two, a belief that came to be known as the Continuum Hypothesis because another name for the real numbers is the continuum. Hilbert gave proving or disproving the Continuum Hypothesis top billing in his list of outstanding problems. Although Cantor and others had tried and failed, it seemed only a matter of time before someone would settle the matter.

In the late 1930s, the Austrian-American logician Kurt Gödel, a close friend of Einstein's at the Institute of Advanced Study, took what appeared to be a significant step towards proving the Continuum Hypothesis. He showed that if it was assumed to be true it didn't contradict the system of nine axioms conventionally taken to form the basis of mathematics – so-called Zermelo–Fraenkel set theory plus the Axiom of Choice (abbreviated collectively to ZFC). But then in 1963, the American mathematician Paul Cohen dropped a bombshell. Cohen showed that *assuming the opposite* – that the Continuum Hypothesis is false – also gave rise to no contradiction with ZFC. In other words, Cohen's work demonstrated, from within ZFC, the Hypothesis is undecidable. In a letter to Cohen, Gödel wrote:

> It is really a delight to read your proof of the inde-
> pendence of the continuum hypothesis. I think
> that in all essential respects you have given the best
> possible proof and this does not happen frequently.
> Reading your proof had a similarly pleasant effect
> on me as seeing a really good play.

The play is not over, however. Mathematicians continue to debate whether the Continuum Hypothesis is 'really' true or false because it seems, in the final analysis, it should be one or the other. After all, we know beyond doubt of the infinity of natural numbers and also of the (larger) infinity of real numbers. How can it be that we can't decide if there's an infinity intermediate in size between these two? Gödel himself argued that the Continuum Hypothesis must surely, in the end, be found to be either true or false. 'Its undecidability from the axioms as known today,' he wrote, 'can only mean that these axioms do not contain a complete description of reality.' The question boils down to what's the most reasonable way to extend ZFC so that the issue is resolved to everyone's satisfaction.

Theoreticians are free to devise any system of axioms they choose but only a system that is consistent, elegant, and, most importantly, useful, will be widely accepted as the bedrock on which to build new and farther-reaching maths. Paul Cohen introduced a technique called forcing, which is a way of expanding the size of a mathematical universe thereby enabling certain matters to be resolved that were previously undecidable. In 2001, the American mathematician and leading set theorist W. Hugh Woodin, at Harvard, proposed adding a new forcing axiom to ZFC that, in the enlarged system, would make the Continuum Hypothesis false. But he's since changed tack, not because of any mistake in the earlier work but because of a new type of axiom that he's devised, known as the inner-model axiom or 'V=ultimate L', which he considers to be more powerful. This new argument reduces some of the philosophical issues surrounding the Continuum Hypothesis to precise mathematical questions that should ultimately be solvable. If Woodin's current line of attack is successful it will lead to the result that Cantor's long-standing conjecture is in fact true and that there's no intermediate infinity between the natural numbers and the reals.

It remains to be seen which of the two main contenders for resolving this problem by augmenting ZFC – with a forcing axiom or with the inner-model axiom – will win. Fans of forcing axioms argue that their approach is the best way of making the foundations of mathematics more useful to traditional branches of the subject. Those favouring the inner model like the idea that being able to prove the Continuum Hypothesis would bring order to the chaos of infinite sets, although it might have little impact elsewhere in maths.

Those working at the cutting edge of set theory are the equivalent of cosmologists or particle theorists in physics. Their research overlaps with metaphysics, ontology, and questions about the ultimate goal of what they're trying to achieve. In venturing into the unknown, mathematicians must decide on the purpose of the axioms upon which they choose to base their explorations and confront the deep nature of mathematics itself. They must ask if it's best to choose axioms for practical expediency or because they're closest to the unadulterated truth of how things are.

Hilbert's second problem also strikes to the heart of mathematical truth and the limits of what can be known. It asks for a proof that the axioms that underpin arithmetic are consistent, in other words, that they don't lead to any contradictions. The arithmetic with which we're all familiar, and that we learn about in school, is known technically as Peano ('piano') arithmetic, after the Italian mathematician Giuseppe Peano. In 1889, he proposed a set of axioms that remains today the generally accepted basis for the maths of natural numbers. Peano's complete system for the natural numbers consists of nine statements, one of which belongs to so-called second-order logic. Peano arithmetic is a weaker system aimed specifically at ordinary arithmetic, concerned with the addition, subtraction, multiplication, and division of numbers. In Peano arithmetic, symbols for addition and multiplication are explicitly included and the second-order axiom replaced by a first-order statement. Hilbert's second problem actually refers to the broader Peano (second-order) scheme but it's nowadays often interpreted as asking if the consistency of Peano arithmetic can be proved.

In 1931, Kurt Gödel shook the mathematical world with the publication of two astonishing theorems. Together his 'incompleteness theorems' showed that in every sufficiently powerful, consistent system of axioms – of which Peano arithmetic is an example – there'll always be statements that can never be proven or disproven, *and the system's own consistency is one of these*. Gödel appeared to scupper Hilbert's hope, expressed in his second problem and later in his grander programme to clarify the foundations of maths, that the consistency of arithmetic could be proven. There's no chance that Gödel's incompleteness theorems will ever be shown to be flawed: they're true for all time. They make it clear that the concept of truth is more powerful than that of proof, which is maddening to the mathematical mind. But the story doesn't end there because, in 1936, the German mathematician-logician Gerhard Gentzen managed to prove the consistency of Peano arithmetic. He did this by working from a different, broader axiomatic system, which, though generally agreed to be itself consistent, would require a still more powerful system to *prove* that consistency. And so it would go on, proving the consistency – the freedom from contradictions – of any axiomatic system always needing a larger mathematical universe to be established. With regard to Hilbert's second problem, then, there are two camps of opinion – the Gödelites and the Gentzenites – just as there are for his first problem.

Hilbert himself, who died in 1943, well aware of these complications and philosophical differences, would have been troubled by them. The indeterminate status of both the first and second problems flew in the face of his entire attitude towards mathematics. He thought that all problems could eventually be solved: it was only a matter of time. In his retirement address to the Society of German Scientists and Physicians in 1930, he famously declared: *Wir müssen wissen. Wir werden wissen.* ('We must know. We shall know.') – the same words that are engraved on his tombstone.

As to which of his 23 problems he most wanted to see solved, Hilbert was clear. 'If I were to awaken after having slept for a thousand years, my first question would be: Has the Riemann Hypothesis been proven?' Named after another German mathematician, Bernhard Riemann, it's the eighth problem on Hilbert's list and widely considered to be the single most important unsolved problem in all of maths. It has to do with the distribution of prime numbers – whole numbers, bigger than 1, that can't be expressed as the product of two smaller numbers. Although there's no pattern or predictability to where individual primes pop up, there *is* some order to how they're distributed en masse. This means we can sensibly ask: given a whole number N, how many primes are there that are smaller than N? In a brief, eight-page paper in 1859, which was his only published work on the subject, Riemann gave the most accurate answer that's theoretically possible to this question – providing that his conjecture is correct. He said, in a nutshell, that the number of prime numbers smaller than N is intimately linked to the 'interesting' or 'nontrivial' solutions of what's become known as the Riemann zeta function, $\zeta(s)$. The solutions are the values of s that make the function equal to zero. Some of these solutions are easy to spot: they happen whenever s is even and negative, and are dismissed as 'trivial'. What Riemann's Hypothesis claims is that all the other solutions – the interesting ones – fall exactly on a single line in something called the complex plane. This is like the ordinary plane we use for plotting x against y except that the horizontal axis represents real numbers while the vertical one represents imaginary numbers – multiples of the square root of -1. Riemann's Hypothesis is that all the interesting solutions of the Riemann zeta function sit on the vertical line in the complex plane that passes through the value ½ on the real number axis. It turns out that where the solutions of the zeta function lie is deeply connected with how often prime numbers show up. In fact, it's possible to write down a formula that tells how many primes there are less than N in terms of the interesting roots of the Riemann zeta function, assuming the Hypothesis is true.

Beyond its role in shedding light on the distribution of primes, the Riemann Hypothesis is important because of the sheer number of different guises in which it pops up in seemingly unrelated areas. 'Assuming the Riemann Hypothesis ...' is the familiar starting point for countless theorems, which would instantly be proved correct if the Hypothesis itself were proved. On the other hand, if just one exception to the Riemann Hypothesis were found, maths would be thrown into chaos. No exceptions have been found among the first trillion or more nontrivial zeroes that have been checked by computer – all lie unwaveringly on the critical line that Riemann predicted they would. In any other branch of science that weight of evidence would be enough to promote a mere hypothesis into a full-blown theory. But not so in maths – and with good reason. Another conjecture about primes stated by Carl Gauss in the mid-1800s was disproved by English mathematician John Littlewood in 1914 and then shown to fail but only above a fantastically large number, known as Skewes' number, equal to 10 to the power 10 to the power 10 to the power 34. Although this known point of failure has since been reduced to about 1.4×10^{316}, it still goes to show that conjectures can hold good up to astronomically large values and then suddenly, surprisingly, break down. No one really expects this to happen with the Riemann Hypothesis but mathematicians won't be happy until an incontrovertible proof or disproof is in the bag.

The Riemann Hypothesis is the only problem that appears in both Hilbert's list and another, compiled exactly a century later, in 2000, by the Clay Mathematics Institute. The fame of this newer list rests not so much on the celebrity of its authors but the fact that a prize of $1,000,000 is on offer for the first verifiable solution to any of the seven problems that it identifies. So far just one of the Clay Millennium Problems – the Poincaré Conjecture – has been solved, though the winner turned down the generous monetary reward on ethical grounds.

The Poincaré Conjecture, named after French mathematician and theoretical physicist Henri Poincaré, is a statement in topology – the study of properties that don't change if a mathematical object is bent or twisted out of shape. At the start of the twentieth century, Poincaré noticed something about loops on surfaces that are finite and have no boundaries, such as the surface of a sphere or a torus (a doughnut shape). Think of a loop as being a curve with the same starting point as ending point, like a circle. Poincaré realised – and proved – that as far as two-dimensional surfaces go, only on a sphere can *any* loop be shrunk to a single point while remaining on the surface. In the case of a torus, for instance, there are loops that go round the hole, which, if you tried to shrink them would end up inside the surface of the torus. Poincaré proposed that this result, concerning loops and spheres, would generalise to higher dimensions. The higher-dimensional equivalent of surfaces (which, by definition, are 2D) are known as manifolds. He noticed that the 3-sphere (the four-dimensional analogue of an ordinary sphere) seemed to be the only manifold in which all loops were contractible. But this time he couldn't prove what became known as the Poincaré Conjecture. Nevertheless he went on to propose the Generalised Poincaré Conjecture: that only on the higher-dimensional counterparts of spheres can any loop be shrunk to a point without leaving the surface. Oddly enough this generalised conjecture turned out to be easier to make progress on than the restricted case for the 3-sphere. In 1960, American mathematician Stephen Smale managed to prove it for all dimensions greater than or equal to five, highlighting a curious phenomenon in topology.

General methods that work in five or more dimensions very often don't work in 3D or 4D. This surprising dichotomy has led to topology in at most four dimensions being known as low-dimensional topology, and topology in five or more dimensions as high-dimensional, as the two fields often use different techniques.

In 1982, American mathematician Michael Freedman managed to solve the Generalised Poincaré Conjecture for 4D, which meant that the Generalised Conjecture had essentially been reduced to the original 3D statement. However, this specific form of the Conjecture turned out to be a much harder nut to crack than any of its higher-dimensional cousins. Important progress was made in 1982 by Richard Hamilton, Davies Professor of Mathematics at Columbia University, in the form of something called Ricci flow, after Italian geometer Gregorio Ricci-Curbastro, upon whose work it was based. He was unable to do more than prove some special cases but, it turned out, Ricci flow was the key to unlocking the Poincaré Conjecture once and for all.

In 2002 and 2003, Russian mathematician Grigori Perelman published three papers that showed how Ricci flow can be used to prove the entire Poincaré Conjecture. There were many gaps in his proof but, unlike with Fermat's Last Theorem, all of these gaps were minor and could be filled using the techniques he described. In 2006, Chinese mathematicians Huai-Dong Cao and Xi-Ping Zhu published a verification of Perelman's proof, but implied that they'd come up with the proof on their own and later had to retract their paper and come clean about the proof being Perelman's. In recognition of his achievement, Perelman was awarded the Fields Medal, considered to be the equivalent of the Nobel Prize for mathematics. However, Perelman refused the Medal and, when awarded the million-dollar Clay Millennium Prize, turned that down as well. His disliked the fame his achievement had brought him and thought it unfair that Hamilton's contribution, which he considered equal to his own, had been overlooked. Never one to seek attention, he became increasingly reclusive until today his whereabouts and activities are something of a mystery.

Among the other Clay Millennium Problems are two that reveal the intimate ties between maths and physics. One – the Yang–Mills and Mass Gap problem – has to do with the world of the very small, the realm in which classical physics gives way to the strange logic and science of quantum mechanics. In 1954, while sharing an office at Brookhaven National Laboratory, Chinese physicist Chen Ning Yang and American physicist Robert Mills hatched a theory to explain the behaviour of the strong force that binds protons and neutrons together in the nucleus. Yang–Mills theory also extends to other ways that subatomic particles can interact, including the electromagnetic and weak forces, and a modern version of it underpins the so-called Standard Model, which is our best theoretical framework for understanding the known fundamental particles. The first part of the Millennium Problem is to come up with a mathematically rigorous quantum version of Yang–Mills that could exist in the real world. The second part is to find the 'mass gap' of this theory, in other words the minimum mass for a particle that it predicts. In the Standard Model, the mass gap is the mass of a glueball, a theoretical particle composed of gluons (the means of holding quarks together in the nucleus), which hasn't yet been observed.

The second of the physics-related Millennium Problems is that old chestnut, the Navier–Stokes Problem. Named after the French engineer Claude-Louis Navier and the British physicist and mathematician George Stokes, the Navier–Stokes equations describe the motion of a fluid while taking into account the pressure and any external forces, such as gravity. Fluids seem to obey these equations, but there is one snag – we don't yet know whether the equations have any solution at all! The major problem is with turbulence, in which the fluid becomes completely chaotic and extremely complicated to analyse mathematically. We do have a 'finite-time blowup' result, where the fluid behaves reasonably for a finite time but then suddenly seems to explode, reaching an infinite distance in a finite time. What we really need is a solution that lasts for all time rather than exploding, and we don't know if that's possible. Once a solution is found, the

Scientists can investigate air flow around objects using a wind tunnel. However, modelling reality with equations becomes difficult, if not impossible, as turbulence sets in.

Navier–Stokes problem goes on to ask whether the solution is 'smooth', in other words, that it avoids any sudden, erratic jumps in the fluid properties.

So how come fluids behave realistically in real life? How is it possible that there could be no solutions to the Navier–Stokes equations, given that in practice fluids don't suddenly explode? The answer is that, like many things in mathematics, the Navier–Stokes equations are merely an approximation to the real world. In reality a fluid is not truly continuous; once you get to a certain level it's made up of individual molecules. The Navier–Stokes equations deal only theoretically with perfectly continuous fluids. Yet it highlights how little we understand about turbulence, even though it's an everyday phenomenon. Indeed, according to one story, when Werner Heisenberg was asked what he would ask God, if given the chance, he replied 'When I meet God, I am going to ask him two questions: Why relativity? And why turbulence? I really believe he will have an answer for the first.'

CHAPTER 14

Could Maths Be Any Different?

> I like mathematics because it is not human and has
> nothing in particular to do with this planet or the whole
> accidental universe – because, like Spinoza's God, it
> won't love us in return.
>
> – Bertrand Russell

IF THERE ARE other intelligent races among the stars, will
their geometry and algebra be the same as ours? If human
history were to be rerun would we inevitably come up with
the same way of doing maths? How much of maths is part
of the fabric of reality, non-negotiable, simply waiting to
be discovered, and how much is of our own invention and
choice?

Anthropologists assume the reason we've adopted a base-
10, or decimal, number system is simply that we have ten
fingers on which to count. In other words, the fact that 10
seems like a 'nice round number' to us is just an accident
of anatomy. If we'd evolved to have eight fingers we'd pre-
sumably count in blocks of eight and have an octal system.
The Yuki people of California and those who speak Pamean
in Mexico *do* have octal systems because they count using

the gaps between fingers rather than the fingers themselves. Maybe octopuses would choose octal, too, if ever they evolved sufficiently to be able to do maths. The Maya and other pre-Columbian cultures of Central America used base-20, perhaps because they counted using all their fingers and toes.

Some animals, like the red panda and the mole, have six fingers on each paw, although the extra digit is really a modified radial sesamoid – one of the bones in the wrist. With six fingers our inclination would be to count in groups of 12 and have a couple of extra digits, for example: 1, 2, 3, 4, 5, 6, 7, 8, 9, Ǝ, ◊. In this case base-12 would seem natural to us, and base-10 unfamiliar and strange.

Members of the Dozenal Society (formerly called the Duodecimal Society) argue that we really should switch to base-12 because it would make calculations a lot easier. The reason for this is that 12 has several factors (other than itself and 1), 2, 3, 4, and 6, whereas 10 has only two, 2 and 5. It would also make telling the time easier, given that there are 12 hours on a clock. Five minutes past two, for instance, would become one and a twelfth hours or 2;1, where ';' is the dozenal equivalent of the decimal point. Ten past two would be 2;2, a quarter past two would be 2;3, and so on.

Although we use a base-10 number system for counting, a huge variety of units have been devised for measuring weight, distance, time, temperature, and other quantities. Those who grew up in Britain in the fifties and sixties will remember having to do arithmetic with a monetary system in which there were not only halfpennies (and, until 1960, farthings, or quarter-pennies), but also 12 pennies to a shilling, and 20 shillings to a pound. School maths exercises became a lot simpler when the UK 'went decimal' on 15 February 1971.

Most countries have adopted not only a decimal currency but also decimal units for measuring other quantities, such as length, mass, and temperature. Elsewhere, especially in the United States and Britain, older units such as pounds, gallons, feet, and miles continue to be widely used, even though working with, say, 12 inches to the foot and 5,280 feet to the mile is more complicated than 100 centimetres to the metre and 1,000 metres to the kilometre. But, of course, although there are different systems of units, the underlying maths – the rules of arithmetic that govern how we do calculations with these units – are the same in all cases.

We may choose to measure distances in feet and inches or metres and centimetres but if we divide the circumference of any circle by its diameter, we'll always get the same value. In base-10 this value is 3.14159..., which translates to 3.11037... in base-8, 10.01021... in base-3, 3.243F6... in base-16 (where F is the base-16 representation of 15 in base-10), and so on. It's a fixed entity in the mathematical universe, so if there were intelligent beings on a planet on the other side of the galaxy, they'd know of this constant, which we call pi, and obtain the same value, although the symbols they used to represent it in any given number base would obviously be different.

The fact that pi is an unchanging fixture of reality – something over which we have no control – didn't deter an attempt to redefine it in law. In 1897, amateur mathematician Edward J. Goodwin tried to convince the Indiana legislature to pass a bill to enact 'a new mathematical truth ... offered as a contribution to education'. Goodwin was convinced, like many cranks before him, that he'd come up with a solution to a classic problem in geometry known as squaring the circle (described in the previous chapter) and was keen that state lawmakers give official backing to his work. One of the effects of this would have been to make pi legally (in this part of the American Midwest) equal to 3.2. The fact that, in 1882, it had been proved beyond a shadow of a doubt that squaring the circle was impossible served as no deterrent to Goodwin. What's more, the Indiana House of Representatives, evidently short of anyone familiar with the Lindemann–Weierstrass theorem (used to finally disprove squaring the circle), was happy to pass the bill. Fortunately, it never became law because by a stroke of good fortune, Professor Clarence Waldo, a mathematician at Purdue University, was in town just before the bill was to be voted on by the state Senate. He was able to enlighten enough senators about the flaws in Goodwin's reasoning and the folly of legislating against mathematical fact that the bill was stopped dead in its tracks.

Pi pops up in a different context at the end of Carl Sagan's novel *Contact* but again in a way that focuses attention on the possibility of meddling with the value of this constant. Ellie Arroway, the SETI researcher who discovers a signal coming from advanced aliens, is eventually told by them about a message that is encoded within the digits of pi. Using a computer program, she finds the message, which starts after about a hundred million trillion places in base-11. Suddenly, the random arrangement of digits that make up pi gives way to a long string of 1's and 0's. The length of the string is the product of two prime numbers. When Ellie uses these numbers to fix the size of a raster and then plots the points (a bright pixel for '1' and a dark one for '0') a very familiar shape appears – a circle! The constant that describes the ratio of a circle's circumference to its diameter contains an encoded picture of a circle within its digits.

The implication is that an incredibly advanced intelligence, possibly present at the origin of the universe, tinkered with the laws of nature so that it could pass on a message within the digits of pi to any beings that evolved to the stage where they could discover it.

Fascinating though Sagan's suggestion is, it contains a flaw: namely that pi is a *mathematical* constant not a physical one. It's true that the geometry of space-time could, in theory, be altered so that actually measuring the ratio of the circumference to the diameter of a circle to great precision would give different values. In fact, we live in a universe that is non-Euclidean because, both locally and over cosmic distances, space-time is curved. But the value of pi isn't determined by measuring the circumference-to-diameter ratio of circles in the real universe. On the contrary, it's the unique value of that ratio for a circle in the space in which the geometry of Euclid applies – perfectly, mathematically flat. Pi also arises in other ways in maths that seemingly have nothing to do with circles, such as the sum of certain infinite sequences (as we saw in Chapter 3). Perhaps Sagan meant to imply that the super-intelligence, which implanted the message into pi, was so far beyond our ken that it could somehow manipulate a constant derived from mathematics itself. This would allow him to suggest that an intelligence might exist that has god-like powers – powers that transcend anything we can understand – without equating it to an actual religious god (Sagan was an atheist). But even gods have to follow the rules of logic and though it's easy enough to imagine other universes in which different sets of physical laws and constants apply, it's hard to see how there could be any tampering with the fundamental nature of mathematics.

Having said that, there is possibly one exception. What if the universe in which we live is not what it seems? What if the universe is not a physical expanse of space and time, and matter and energy, but instead is a simulation? This disturbing scenario has been much discussed in recent years by philosophers and even some scientists. Today's high-speed computers and sophisticated software can already generate simulated worlds with which we can interact as avatars and explore a realistic but entirely fictional landscape. These simulated worlds (in video games) obey different sets of rules, designed to give the player a novel and exciting experience. Nevertheless, the rules are consistent and make sense within the system of which they're part.

As immersive technology evolves and devices such as neural interfaces become more effective and readily available, we'll be able to disappear for hours at a time into an entirely computer-generated alternative world, which will seem as tangible and convincing as reality itself. But what if 'reality itself' is a simulation and we ourselves and everything around us are mere artefacts in an alien computer of awesome speed and power? There would then be no limit to how much we and our fabricated universe could be manipulated from the outside. It would be quite possible to implant patterns or messages into irrational numbers such as pi because the values of such numbers could be made part of the simulation and could be controlled externally at will. What we regard, on the one hand, as the laws of physics, and on the other, the immutable Platonic realm of mathematics, may both be arbitrary constructs of some fantastically powerful computer program.

Assuming, though, that we're not unwitting digital denizens in some elaborate fantasy – that we're flesh-and-blood beings in an honest-to-goodness, physical universe – how different could maths be? Suppose we could reset time to the start of human civilisation and let history run again but along a new course with a different set of circumstances and central cast members. Inevitably, discoveries would be made in a different order, and in different places and times, and there'd be more development in some areas of maths and less in others than occurred on our timeline. Perhaps the Greeks would have invented algebra and not given so much attention to geometry. The ideas of set theory and Cantor's work on infinity might have occurred to some genius in Renaissance times or in ancient India.

A glimpse of how such variations might have affected the appearance of maths was offered by a short-lived but major change in the way maths was taught in American grade schools in the 1960s. So-called New Math was introduced in an attempt to boost science and maths skills in the wake of the Soviet Union's shock successes in the Space Race, beginning with the launch of Sputnik 1 in 1957. Suddenly, in place of traditional arithmetic, children were expected to learn about modular arithmetic (in which numbers wrap around after reaching a certain value), bases other than 10, symbolic logic, and Boolean algebra. Such concepts proved baffling not only to young students, more familiar with learning times tables and number bonds, but also to their teachers and parents. Many parents, in fact, began sitting in on their sons' and daughters' classes in order that they might be able to help their children with homework.

Through New Math it was hoped that a generation would grow up able to accelerate America's progress in technology, especially in areas such as electronics and computers, so as to outpace the Soviets. The great weakness of the new scheme, though, as soon became clear, was that it expected children to make a mental leap to abstract topics and methods with which they were wholly unfamiliar. American mathematician and author of several widely used textbooks George Simmons wrote that New Math produced students who had 'heard of the commutative law but did not know the multiplication table'.

As an experiment in education, New Math failed and was soon abandoned. However, it did give an interesting insight into how maths can appear very different if presented in a wholly novel fashion. What's more, just because New Math didn't work out as planned doesn't mean that young children can't assimilate concepts that we don't normally meet until we're much older, if at all. One of us (David) has been privately tutoring children of all ages, from 5 to 18, for several decades, and has found that even youngsters who've just started primary school can begin to grasp ideas such as infinity, higher dimensions, and unusual geometries (like that of the one-sided Möbius strip) if they're introduced in simple language and in a way that's entertaining and fun. In fact, he's convinced that people could come to have a deep and intuitive appreciation of such exotica as the fourth dimension and transfinite numbers if they were encouraged to play and engage with such things from an early age. It's much the same as with languages. Young children who grow up in bilingual environments have no trouble absorbing and becoming fluent in, say, English and Spanish, whereas learning a second language as an adolescent or adult is generally much harder.

So, it's clear that maths could look a lot different from how it does if history had taken a different tack. We might tend to think more in terms of shapes than numbers, or (as

The Lovell Telescope at Jodrell Bank, Cheshire, which has been used to search for signals from extraterrestrial intelligence.

was attempted with New Math) be more comfortable with set theory than ordinary arithmetic and algebra. Such differences might be even greater on other worlds where biological evolution leads to life forms that are completely foreign to anything we see on Earth.

In his 1961 novel *Solaris*, Polish writer Stanislaw Lem imagined a planet on which there was a kind of thinking sea – a single, unbroken planet-wide intelligence. So utterly alien does it prove to be to the human explorers who try to interact with it from an orbiting spacecraft that all attempts at meaningful communication break down. What kind of maths would such a being come up with? It seems at least possible that an organism with no concept of other individuals, or of separate things in its environment, would not go down the route that we did of learning how to count, and doing other simple arithmetic, using natural numbers. Such a being would be much more likely to think in terms of continuous quantities rather than discrete ones. It might start out then by developing the mathematics of smooth functions and only much later discover whole numbers and how to work with them. Whether single planet-wide life forms like this actually exist somewhere we have no idea, but just thinking about the possibility suggests that, under other circumstances, maths could progress along radically different lines. There's nothing to say that wherever maths arises it has to start out with what we consider to be basics like integers and Euclid's geometry. The appearance of extraterrestrial mathematics could be very unusual indeed. Nevertheless, the parts of maths explored and established by the human race should correspond exactly with those same parts encountered and investigated by other intelligent races in space. Our art, music, languages, and technology might differ enormously but the fundamentals of maths should be the same everywhere.

Where we might see significant discrepancies is in the basic assumptions made on which systems of mathematics are built. These basic assumptions, known as axioms, are the bedrock on which all our theorems and proofs rest. At the dawn of recorded history, when people first started to use numbers and develop rules of thumb for working with shapes, areas, and the like, they just did what was useful from a practical standpoint. The first person, as far as we know, to think long and hard about the logical foundation of maths was Euclid in about 300 BCE. The results and proofs that he arrived at in his great work on geometry, the *Elements*, were built on a set of five postulates (roughly equivalent to what we'd now call axioms) and five more statements that he referred to as 'common notions'. Among the postulates are that a straight line can be drawn from any point to any point and that all right angles are equal to one another. They all seem obvious and uncontroversial except for one – the fifth, which is also known as the parallel postulate. Euclid's statement of the parallel postulate is pretty long-winded and doesn't specifically mention parallel lines but it's equivalent to the statement: 'two lines that are parallel to the same line are also parallel to each other'.

Even the ancient Greeks weren't as happy about this fifth postulate as they were with the other four. It was more complicated and less self-evident. The fact that it comes last in Euclid's list of postulates and that he didn't use it at all in deriving his first 28 theorems suggests that he felt there was something slightly unsafe with including it as a core assumption. Yet he recognised that he needed it in order to move ahead with his system of geometry – what we now refer to as Euclidean geometry. Over time, many mathematicians tried to derive the fifth postulate from the other four – and, in every case, failed. The first person to see clearly where the problem lay was the German mathematician Carl Gauss. He started his investigation of the foundations of Euclidean geometry when he was just fifteen but it took him another quarter of a century to become convinced that the parallel postulate was independent of the other four. At that point he began to look at the consequences of leaving out the fifth postulate altogether and, in so doing, caught the first glimpse of a strange new geometry. In a letter to a colleague, he wrote:

> The theorems of this geometry appear to be para-
> doxical and, to the uninitiated, absurd, but calm,

steady reflection reveals that they contain nothing
impossible …

Not being one to court controversy, Gauss didn't publish his
findings although he considered doing so towards the end
of his life. It was left to others, including one of his friends,
the Hungarian mathematician János Bolyai, and the Russian
Nikolai Lobachevsky, to bring non-Euclidean geometry to
the attention of the world.

The discovery that there exist other forms of geometry
beyond that formulated by Euclid doesn't disprove Euclidean
geometry. What it shows is that starting from a different
collection of axioms, different systems of mathematics,
each consistent within itself, can be built up. We're free to
choose these axioms at the outset, providing they don't contradict each other, and then deduce theorems and produce
proofs based on them. Naturally, when mathematicians go
about their business they try to choose starting assumptions
that seem to make sense and that serve some useful end.
A set of axioms developed by the German mathematician
Ernst Zermelo and the German-born Israeli mathematician
Abraham Fraenkel in the first quarter of the twentieth century (with the addition of something called the Axiom of
Choice) is currently accepted as the most common basis of
mathematics. But it doesn't have to be that way. Our maths
could be founded on any number of different collections of
core premises.

A lot of the axioms we choose to develop in mathematics
are geared to fit our intuition – human intuition. Another
race of sentient beings, whose physical experience contrasted
greatly with ours, might start out with radically different
axioms and end up with a very alien-looking system of

maths. That doesn't alter the fact that if we began with those same alien axioms we would arrive at exactly the same alien-looking system. To the best of our understanding, mathematics is universal. It may be that elsewhere it has developed in a different order and along very different lines, but granted the same set of starting assumptions and rules it must inevitably arrive at the same theories and conclusions.

Acknowledgements

ONCE AGAIN WE'RE tremendously grateful to our families for all their love, patience, and support. Our thanks also go to the staff at Oneworld, especially our editor, Sam Carter, and assistant editor, Jonathan Bentley-Smith, who've helped make the *Weird Maths* series so much fun to work on.

For more weird maths, please visit
weirdmaths.com